Ways to Go Beyond and Why They Work

Rupert Sheldrake

Ways to Go Beyond
and
Why They Work

Spiritual Practices in a Scientific Age

CORONET

First published in Great Britain in 2019 by Coronet
An Imprint of Hodder & Stoughton
An Hachette UK company

This paperback edition published in 2020

5

Copyright © Rupert Sheldrake 2019

The right of Rupert Sheldrake to be identified as the Author
of the Work has been asserted by him in accordance with the
Copyright, Designs and Patents Act 1988.

A CIP catalogue record for this title is available from the British Library

B format ISBN 9781473653443

Typeset in Sabon MT by Palimpsest Book Production Limited,
Falkirk, Stirlingshire

Printed and bound in Great Britain by Clays Ltd, Elcograf S.p.A.

Hodder & Stoughton policy is to use papers that are natural,
renewable and recyclable products and made from wood grown
in sustainable forests. The logging and manufacturing processes
are expected to conform to the environmental regulations
of the country of origin.

Hodder & Stoughton Ltd
Carmelite House
50 Victoria Embankment
London EC4Y 0DZ

For Rick Ingrasci and Peggy Taylor,
friends and inspirers for 35 years

Contents

Preface

I am a strong believer in the scientific method and empirical enquiry. I am a research scientist, and have done hundreds of experiments myself, summarised in more than eighty-five papers in peer-reviewed scientific journals, together with articles in many technical books and in several scientific encyclopaedias. I spend much of my time doing experimental research.

I also believe that it is only through spiritual practices and direct experiences that we can effectively deepen our own connections with the more-than-human realms of consciousness, and become more aware of the underlying source of all consciousness and all nature.

These practices tell us something about our own nature. They also tell us something about the nature of the spiritual realm, the realm beyond the mundane, by which I mean the realm of more-than-human consciousness.

This is a sequel to my book *Science and Spiritual Practices*. In that book, I discuss seven very different spiritual practices – gratitude, meditation, connecting with nature, relating to plants, singing and chanting, rituals, and pilgrimage. In this book, I discuss a further seven practices, including some that are not usually thought of as having a spiritual dimension, like learning from animals and participating in sports.

The scientific exploration of spiritual practices is a very positive aspect of the modern world. The self-imposed separation between science and the spiritual realm is breaking down, as scientists investigate spiritual practices and as the field of consciousness studies develops. Both science and spiritual experiences are empirical, based on experience. Systematic research into experiences brought about by spiritual practices brings science and spirituality into convergence. This new synergy could lead towards better ways of relating to the realm of more-than-human consciousness, and also to deepening our understanding of spiritual experiences.

In the preface to *Science and Spiritual Practices*, I wrote about my own personal background and long-standing interests in both scientific research and the exploration of spiritual practices, not only in Europe and North America, but also in Asia. In summary, after a conventional Christian upbringing, as a teenager studying science I became an atheist, accepting the conventional scientific view that science was essentially atheistic and that science and reason had superseded superstitions, like a belief in God and in religious dogmas.

But through my studies at Cambridge and Harvard Universities, and through my research in developmental biology at Cambridge, my atheist faith was weakened by scientific doubts about the machine theory of life. I found it harder and harder to believe that animals and plants are nothing but unconscious mechanisms programmed by genes that had evolved by chance and the blind forces of natural selection.

While I was a Fellow of Clare College, Cambridge, and Director of Studies in Cell Biology, I also found myself questioning the materialist assumption that minds are nothing but the activity of brains. Travelling in Asia, discovering psychedelics, and taking up the practice of meditation amplified these doubts.

I spent a year in Malaysia doing research on tropical plants at the University of Malaya. I also spent seven years in India, five of them as Principal Plant Physiologist and Consultant Plant Physiologist at the International Crops Research Institute for the Semi-Arid Tropics (ICRISAT), near Hyderabad, where I helped in the breeding of new varieties of chickpeas and pigeon peas, and in the development of new cropping systems that are now widely used by farmers. I also spent two years in a Christian ashram on the bank of the river Cauvery in Tamil Nadu, where I wrote my first book, *A New Science of Life*.

My wife, Jill Purce, is a voice teacher, who has pioneered new ways of working with group chant. Her workshops and seminars open up ancient spiritual practices with the voice to anyone who is interested. Jill's work has shown me over and over again how spiritual practices can enrich people's lives, whether they think of themselves as religious or not, and is one of the inspirations for this book.

My purpose in this book is not to make an exhaustive catalogue of all possible spiritual practices, but to illustrate that there are many different ways of approaching the spiritual realm, and that they have scientifically measurable effects.

I hope this book will help those who already do these spiritual practices to see them in a new evolutionary and scientific light. For those who are not familiar with some or all of these practices, I hope these discussions will open up new personal possibilities.

Hampstead, London
July 2018

Introduction

Never before has any civilisation had access to almost all the world's spiritual practices. In major cosmopolitan cities, it is now possible to attend rituals from a wide range of religious traditions, to learn to meditate, to practise yoga or chi gong, to take part in shamanic practices, to explore consciousness through psychedelic drugs (albeit illegally in most places), to sing and chant, to participate in a wide range of prayers, to learn martial arts and to practise a bewildering array of sports.

All these practices can take us beyond normal, familiar, everyday states of consciousness. They can lead to experiences of connection with more-than-human consciousness, and a sense of a greater conscious presence. Such experiences are often described as spiritual.

The experiences themselves leave open the question of the nature of the spiritual realm. As I discuss in this book, there are several possible interpretations of spiritual experiences, including the materialist view that they are all inside brains and that there are no more-than-human forms of consciousness 'out there'.

At the same time, spiritual practices are being investigated scientifically as never before. We are at the beginning of a new phase of scientific, philosophical and spiritual development.

The philosophy of nature that has dominated the natural sciences since the late nineteenth century is mechanistic materialism, the belief that all nature is machine-like, made up of non-conscious matter whose behaviour is determined by mathematical laws, including the probabilistic laws of quantum mechanics, together with chance events. Nature is purposeless, and evolution has no direction or meaning. Minds are nothing but the activity of brains, and are confined to the insides of heads. Consciousness is an illusion, or a functionless by-product of brain activity. Free will is an illusion, too.[1]

This kind of science has been very successful in physics, chemistry, engineering and technology. It has also led to great advances in biology, in particular a description of the molecular mechanisms of genetic inheritance and the role of genes in the synthesis of protein molecules. In medicine, there have been enormous advances, through ever more skilful surgery and through the development of a host of drugs and vaccines.

But the science of the mind, psychology, has been far less successful. A science that assumed that everything is made of non-conscious matter was not well-equipped to deal with consciousness. For much of the twentieth century, academic psychology in the English-speaking world was dominated by the school of behaviourism, which treated subjective experiences as irrelevant and unscientific, because they could not be objectively measured. Behaviourists valued only objective measurements of bodily activity or glandular secretions.

When behaviourism's dominance waned in the late twentieth century, the new academic orthodoxy was cognitive psychology, based on the idea that the mind is nothing but the activity of the brain, and that the brain is a kind of computer, processing algorithms. The mind is like the software and the brain is like the hardware.

In the 1990s, the psychologist Antonio Damasio helped disrupt the dominant computer-centred model of brain activity by suggesting that thinking is influenced by emotions, and that emotions are rooted in the reactions of the body, like the fight-or-flight reaction mediated by the hormone adrenaline. In his influential book, *Descartes' Error: Emotion, Reason and the*

Human Brain, he proposed that emotions guide behaviour and decision-making.[2] For most people outside the academic world, the idea that emotions influence human behaviour and decision-making is neither shocking nor original. It is common sense. But in the world of academic psychology it was a radical innovation.

Another innovation was the school of positive psychology, officially established in the academic world in 1998, which set itself the goal of finding out what makes life worth living.[3] When positive psychologists investigated what makes people happy, they found a wide range of activities that do so, including being absorbed in one's work, having a good conversation, singing, dancing, making love and playing games. The common factor is a state of absorption or flow.[4] By contrast, people are less happy when they are separate, detached, alienated, or in conflict. Positive psychologists have contributed to the study of spiritual practices by studying the effects of gratitude (discussed in *Science and Spiritual Practices* [*SSP*], chapter 2), and the induction of states of flow through sports, singing and dancing.

The 1990s also saw the emergence of the field of consciousness studies, which went far beyond the limited perspectives of behaviourism and cognitive psychology, opening up areas of research that most scientists had ignored or rejected as unscientific. Researchers in this field investigate near-death experiences, meditation, the effects of connecting with nature, out-of-the-body experiences, mystical states of union, the effects of singing and chanting in a group, and psychedelic experiences. All these enquiries take the sciences a long way beyond old-style mechanistic materialism.[5] Indeed, some materialists and atheists are now exploring their own consciousness through practices like meditation and psychedelics. Sam Harris, for example, one of the New Atheists, is now a meditator and teaches meditation online.[6]

This convergence of science and spiritual practices is surprising from the point of view of materialist orthodoxy, in which the vast majority of contemporary scientists have been trained. Yet it is entirely consistent with the scientific method, which involves the formation of hypotheses – guesses about the way the world works – and then testing them experimentally. The ultimate arbiter is experience, not theory. In French, the word *experience*, means both 'experience' and 'experiment'. The Greek word for experience is *empeiria*, the root of our English word 'empirical'. The exploration of consciousness through consciousness itself is literally empirical, based on experience. Spiritual practices provide ways in which consciousness can be explored empirically.[7]

This book continues the enquiry I started in my book *Science and Spiritual Practices,* in which I discuss seven different practices that have been investigated empirically, both by the practitioners themselves and by scientists studying the effects of their practices. The chapters discussing these practices are entitled:

1. Meditation and the Nature of Minds
2. The Flow of Gratitude
3. Reconnecting with the More-Than-Human World
4. Relating to Plants
5. Rituals and the Presence of the Past
6. Singing, Chanting and the Power of Music
7. Pilgrimages and Holy Places

In this book, I discuss seven further practices under the chapter headings:

1. The Spiritual Side of Sports
2. Learning from Animals
3. Fasting
4. Cannabis, Psychedelics and Spiritual Openings
5. Powers of Prayer

I conclude with a discussion of why these practices work.

These two books do not constitute a comprehensive survey of all spiritual practices. There are many others, including yoga, service to others, tai chi, chi gong, devotional worship or *bhakti*, tantric sex, caring for dying people, dream yoga, and the practices of the arts. Some practices suit some people better than others; some practices are better at some times of life than others, and all religious traditions have their own combinations.

I have taken part in all the practices I discuss in this book. Many people are more expert than I am in each of these practices. I am not a guru, but an explorer. My purpose is to take further the project I began in *Science and Spiritual Practices*, to show that there is a wide variety of ways to connect to greater conscious realities, however we conceive of them, and that the effects of these practices can be investigated scientifically. At the end of each chapter, after the discussion of a particular kind of practice, I suggest two ways in which readers can experience this practice for themselves.

We are on the threshold of a new era of the exploration of consciousness, both through a revival of spiritual practices and also through the scientific study of them. After several generations in which science and spirituality seemed to be in opposition, they are becoming complementary. Together, they are contributing to an unprecedented phase of spiritual evolution, beginning now.

I

The Spiritual Side of Sports

Most people do not think of sports as spiritual practices; sports seem supremely secular. Yet in modern secular societies, sports may be one of the most common ways in which people experience the self-transcendence that can come through being in the present. A meditator may find his mind wandering and only occasionally come back into a full sense of presence, but a football player in an important match is completely in the present, or else he is out of the game. Someone skiing downhill at sixty miles an hour has to be completely focused, as does a surfer on a gigantic wave, or a free climber on a rock face with no ropes, or a hunter stalking a deer when the slightest noise or visible movement might cause the quarry to run away.

Surprisingly, the word sport is indirectly derived from the Latin root *portare*, meaning to carry, as in our English words export (to carry out), deport (to carry away) and disport (away from carrying) – which came to mean to amuse oneself, or to make merry, or to play games. Sport comes from disport.

Play comes from the Old English *plega*, to frolic.[1] The primary English meanings are to exercise oneself, or to act or move energetically. Play also means to engage in a game, or to play for stakes as in gambling, or to take part in a sport, or to play a musical instrument, or to perform on the stage, as in a play. Not only humans and non-human animals play: flames and fountains play too, metaphorically, through their freedom and spontaneity, as does moving light in 'the play of light'.[2]

The Middle English word *gamen*, related to the Old High

German *gaman*, merriment, meant a game or a sport, and came to mean gaming in the sense of gambling. Game also means wild animals caught for sport, as in 'game pie'.

All these words have a wide range of meanings, but what they have in common is being away from the usual business of life. The philosopher David Papineau, himself a keen sportsman, has thought more about sports than most people, and has summarised his conclusions with admirable clarity: the value of sporting achievement lies in 'the enjoyment of sheer physical skill'.[3]

Humans hone their physical abilities and take delight in exercising them. This definition explains why many sports are not games, like skiing or shooting pheasants, while some sporting skills exist only within games, like topspin tennis backhands. Other sports are based on skills that already occur in everyday life, like running, jumping, rowing, shooting, lifting and throwing.

Papineau concludes, 'These ordinary activities turn into sports whenever people start performing them for their own sake, and strive for excellence in their exercise.'[4] A wide range of other physical activities that are not part of everyday life can also turn into sports, like windsurfing and skydiving.

What about the role of competition? Some sports, especially spectator sports, are competitions, like wrestling and cricket. Papineau points out that competition plays an important part in sport, because it enables people to measure themselves against others: 'To exercise a skill is to want to do something *well*, indeed as well as is feasible.'[5] But some sports are not directly competitive. Mountaineers may seek to scale a particularly difficult peak, but their achievement is not primarily in competition with other people, more a challenge to themselves.

In this chapter, I look first at the evolutionary and anthropological context of modern sports. Although sports are not normally undertaken as spiritual exercises, they can have a range of spiritual effects, which I discuss. These effects include being intensely present, and feeling part of something greater than oneself.

The evolutionary background

In his book, *The Descent of Man*, Charles Darwin emphasised the importance of sexual selection in a wide range of animals. Among birds, he pointed out that most males are 'highly pugnacious' in the breeding season, but even the most aggressive rarely depend solely on their ability to drive away rivals; they have a range of ways of 'charming the females'. Some depend on the power of song, others on courtship dances, and yet others on 'ornaments of many kinds, the most brilliant tints, combs and wattles, beautiful plumes, elongated feathers, top-knots, and so forth.'[6]

By contrast, among mammals, force is more important: 'The male appears to win the female much more through the law of battle than through the display of his charms. The most timid animals, not provided with any special weapons for fighting, engage in desperate conflicts during the season of love . . . male squirrels engage in frequent contests, and often wound each other severely, as do male beavers, so that hardly a skin is without scars.'[7]

In most mammalian species, males are larger than females; and the difference is greatest in species where males fight most. In a polygamous seal species, the males are about six times heavier than the females, while in monogamous species there is little difference between the sexes. Similarly, male right whales, which do not fight each other, are the same size as females; male sperm whales, which fight for mates, are double the size of females.

Darwin also showed that in many mammalian species, only males have tusks or horns, and they use them for fighting and defence. In some species, females also have tusks or horns, but smaller and less developed than in males. Darwin concluded they had evolved primarily in the males in the context of sexual selection and then carried over to females secondarily. Similarly, defensive features are much more pronounced in males than females, as in the manes of male lions, which protect their necks from being bitten by other male lions in fights.

It was a short step for Darwin to point out that similar forms of sexual selection are widespread in human societies. He quoted a report by Samuel Hearne, who explored remote parts of Northern Canada in the late eighteenth century. Hearne wrote:

> It has ever been the custom among these people for the men to wrestle for any woman to whom they are attached; and, of course, the strongest party always carries off the prize. A weak man, unless he be a good hunter, and well beloved, is seldom permitted to keep a wife that a stronger man thinks worth his notice. This custom prevails throughout all the tribes, and causes a great spirit of emulation among their youth, who are upon all occasions, from their childhood, trying their strength and skill in wrestling.[8]

Where Darwin led, twenty-first-century evolutionary psychologists followed, seeing sports as 'culturally invented courtship rituals'.[9] Many studies have shown that levels of the hormone testosterone increase before competitive sports in both men and women, and in men these levels increase further following victory, and decrease following defeat.[10] This effect occurs not only in participants, but also in male spectators.[11] The performance-enhancing effects of testosterone have been known since the 1930s, and it has been widely used to improve performance in sports events. It is now prohibited in competitions, but this rule is hard to enforce through blood tests, because natural levels of testosterone are so variable.[12]

The ancient evolutionary roots of competition in sports may also play a part in the well-established phenomenon of the 'home advantage', whereby teams in a range of sports consistently perform better when playing at home, as if defending their own territory. Measurements of testosterone in saliva samples of soccer players showed that the levels were higher before home than away games, and were also higher before playing an extreme rival than a moderate rival.[13]

If sports are elaborate mating rituals, then why do so many people enjoy watching members of their own sex compete? One answer suggested by evolutionary psychologists is that this enables spectators to assess the players as cooperative partners and as prospective leaders.[14] But this argument cannot account for watching fights between other species, such as cockfights, which were a popular spectator sport in several ancient cultures, including India, China and Persia, and are still widespread in many parts of the world, especially South-East Asia, although they are now illegal in many countries because of their cruelty.

These evolutionary perspectives apply most obviously to competition between individuals, as in wrestling, boxing, martial arts, fencing, racing, archery, javelin throwing and tennis. The same principles apply to competition in hunting, shooting and fishing. By contrast, in team games there is not only competition between the teams, but cooperation within them. In some cultures, such games were not far removed from wars, as discussed below.

The anthropology of sports

Many traditional sports are based in male rivalry, but there are great cultural differences in the sports people play.

The most common kind of contest all around the world is fighting between young men. For example, among the Yahgan, a small group of hunter-gatherers living at the southernmost tip of South America, wrestling was the most popular sport. Before the match began, a young man chose his opponent from the audience and challenged him by placing a little ball at his feet. This man had to accept, if he did not want to be called a coward. Then the two men stepped inside the large circle of spectators. They wrestled until one held the other down on his back.

The defeated man suffered an affront to his honour and that of his friends, so one of his supporters immediately stepped into the circle, grasped the ball and placed it at the feet of the winner, who then had to prove his skill with this new opponent.

Girls preferred successful wrestlers as marriage partners. Because it was so important, a loser was given several opportunities to save face. If the winner continued winning, the match could descend into a general mêlée, so no one was a clear winner, salvaging reputations.[15]

As well as fighting sports, ball games are common in many parts of the world. Among the Kurnai people in Australia, the ball was the scrotum of a kangaroo tightly stuffed with grass.[16] In the cultures of Mesoamerica, ball games were an important part of the culture, but here the balls were generally made of rubber and the games took place in courts. The archaeological evidence suggests these games were being played at least as early as 1500 BC, and ball courts are an important part of Mayan archaeological sites. These were not just games, but complex rituals in which the movement of the ball symbolised the movement of the heavenly bodies across the skies, and the mythology of the game was filled with struggles between gods.[17] There was extravagant betting on these ball games, and the wagered items included jewellery, slaves, women, children and even entire kingdoms.

Among the Aztecs and Mayans, the outcome of the game for some players was death. At the Mayan site at Chichén Itzá, in Mexico, a bas-relief on the wall of the large ball court seems to depict the captain of a losing team being sacrificed. The heads of victims were displayed on racks. The ball game was a mock war and could even function as a substitute for war.[18]

One of the few social groups studied by anthropologists that did not play games were the Dani, who live in the highlands of western New Guinea. Their ritual life was much concerned with propitiating and appeasing ghosts and spirits to guard against unwanted intrusions. Although the Dani did not play games, they were great fighters, and one of their motives was to avenge the death of an ancestor, because the spirit of the ancestor demanded it. For every person killed, someone had to pay; to balance the account, another life had to be taken. But the account was never balanced, and so the battles went on and on.

These sport-like contests were generally staged on the open plains of a valley floor, fought with wooden spears and bows and arrows, and surrounded with much pomp and enthusiasm. They took place according to a set of rules and were marked by a playful attitude that took precedence over the idea that someone must be killed.

These battles were an example of what anthropologists call 'social warfare'. Social wars, as opposed to economic wars, are not fought for taking over territory, capturing resources or subjugating people. They are about prestige, honour, revenge and entertainment. When this system prevailed, more than a quarter of the deaths of young men among the Dani were the result of warfare. Nevertheless, young men found battles exhilarating. According to an anthropological observer:

> There is a tremendous amount of shouting, whooping and joking. Most men know the individuals on the other side, and the words which fly back and forth can be quite personal. One time, late in the afternoon, a battle had more or less run out of steam. No one was really interested in fighting anymore and some men began to head for home. Others sat around on rocks and took turns shouting taunts and insults back and forth across the lines, and connoisseurs on both sides would laugh heartily when a particularly witty line hit home.[19]

There are many traditions of sacred games. In ancient Greece, the Olympic Games were held every four years in honour of the sky god Zeus, at the village of Olympia. These were not simply games; they sacralised the fundamental principles of rule-bound contests, namely fairness and playing by the rules. The philosopher Plato – whose name meant 'broad shouldered' – was himself a wrestler, and advocated the cultivation of physical virtues for the building of character.[20] In classical Rome, the sacred games were considered essential for the wellbeing of the state.[21]

The modern Olympic Games are based on these ancient models, and although they are conducted in a secular spirit, they provide an incentive that causes many thousands of people to train and aspire to their highest abilities. Some, impelled by the competitive spirit, illicitly take anabolic steroids or other chemical aids to spur them onto higher levels of achievement. And many pray and ask for God's blessings, or the blessings of their patron saints or deities.

In feudal China, war was treated as if it were a game, like chess being played in real life. Warlords conducted their battles sitting high above the battle itself, with each lord ordering the field manoeuvres of his soldiers in terms of moves. At a set time, the battles stopped and troop positions were marked, so that the battle could resume at the same point on the following day. The rival warlords then gathered together for refreshments, discussed the day's successes and failures, and compared the performance of their troops.[22]

In many traditional societies, sports played a major role in exercising and refining skills necessary for warfare. Among the Zulus in South Africa, for example, one of the favourite pursuits of boys was sham fighting with sticks. Every boy, like every man, when away from his home carried a couple of strong sticks. One stick, held about the middle in the left hand, was used for parrying and the other for striking.

When soccer was introduced to Zululand, it soon became a

major leisure-time activity and many of the same principles applied to football games as to warfare. The Zulu football teams now employ their own witch doctors, who do magic to help their own team, and cast spells to weaken their opponents, as if they were going to war. The teams also apply a strict discipline. They camp out on the nights preceding matches, and undergo purification rites similar to those once performed before battle. The teams move onto the playing field in military formation.[23]

In the United States, the British game of rugby has been transformed into American football, which is even more violent than rugby itself. Dave Meggyesy, a college footballer at Syracuse University and then a professional player with the St Louis Cardinals, has written a fascinating book about his experiences, *Out of Their League.*

He describes a game in which he was playing for the Cardinals against the Pittsburgh Steelers. His assignment was to block one of the men on the Pittsburgh team: 'Running full speed and not looking to either side. I knew he didn't see me and I decided to take him low. I gathered all my force and hit him. As I did, I heard his knee explode in my ear, a jagged, tearing sound of muscles and ligaments separating. The next thing I knew, time was called and he was writhing in pain on the field. They carried him off on a stretcher and I felt sorry – but at the same time, I knew it was a tremendous block and that was what I got paid for.'[24]

Meggyesy excelled at football at his high school in Ohio, and as the captain of his team he had to give a speech to encourage the players: 'At the beginning of the game, I would give my "We've got to get them, we've got to get them" speech. We were really fired up and felt we were going to annihilate "them". I particularly didn't want to see their faces, because the more anonymous they were, the better it was for me – and I'm sure most of the other ball players felt the same way: They were a faceless enemy we had to meet.'

When he was in his college team, he had a teacher who 'had the quality of a killer about him' and who explained his philosophy as follows: 'In football the commies are one side of this ball and we are on the other. That's what this game is all about, make no mistake about it.'[25] Sometimes, Meggyesy wrote, 'I felt great pleasure and release from the sheer physical violence of the game. Sometimes after getting a clean shot of the ball carrier, I would feel this tremendous energy flow and not experience the pain of contact at all. I sometimes could psych myself so high, I would feel indestructible.'[26]

Meggyesy sustained some serious injuries, but was generally expected to play on in spite of them, if necessary with injections of a local anaesthetic to reduce the pain. And there was a strong incentive to minimise the importance of injuries. He found that one of the worst things that could happen to a player was to get a serious injury in training. He would fall behind everyone else in learning and practising and would be ostracised by the coaching staff. 'Healthy ball players don't like to fraternise with an injured man either. It's like some voodoo in which the injured player becomes a sort of leper.'

Sports are now a multibillion-dollar business and professional players are no longer just disporting themselves. They have serious and often well-paid jobs. The intrinsic joy of physicality can easily give way to the desire for approval, money, fame and adulation.[27] Yet although commercialisation has transformed sports, many millions of people still play as amateurs, and hundreds of millions of people participate in sports vicariously as spectators, both through attending the events themselves or watching them on television. Being a part of a crowd of supporters, united in support of a team, can be a powerfully bonding experience, especially when enhanced by singing together – as in the singing of 'You'll Never Walk Alone' before matches by thousands of Liverpool Football Club supporters.[28]

The pioneering work of Michael Murphy

Michael Murphy co-founded the Esalen Institute in California in 1962 and was a pioneer of the human potential movement. He was among the first people to point out that sports are one of the most common ways in the modern world in which people experience altered states of consciousness, and even mystical experiences. His novel, *Golf in the Kingdom,* first published in 1972, was an international bestseller, and tells the story of a young man, Michael Murphy himself, who stopped in Scotland on the way from California to India.

His life was changed when he played a round of golf with a mystical professional golfer called Shivas Irons, who explained:

> The game requires us to join ourselves to the weather, to know the subtle energies that change each day upon the links and the subtle feelin's of those around us. It rewards us when we bring them all together, our bodies and our minds, our feelin's and our fantasies – rewards us when we do them and treats us badly when we don't. The game is a mighty teacher . . . The grace that comes from such a discipline, the extra feel in the hands, the extra strength and knowin', all those special powers ye've felt from time to time, begin to enter our lives.[29]

In real life, Murphy spent a year and a half at the ashram of Sri Aurobindo in Pondicherry, India, and was much influenced by Aurobindo's evolutionary philosophy, which emphasised not only the spiritual evolution of humanity, but also its physical evolution as part of what he called Integral Yoga. Murphy, himself a keen golfer and marathon runner, came to see that sports are 'the yoga of the West', and his vision of human potential included the realisation of new potentials through sports:

In no other field of human activity is there such a proliferation of specialised physiques. For as athletics have developed in the modern world, they have required an ever greater variety of skills and body structure to support them – whether it is the muscular frame of a 270-pound defensive tackler [in American football], the elastic joints of a gymnast, the prodigious cardiopulmonary system of a marathon runner, or the steady hand of an archer. Never before have there been so many experiments with the body's limits.[30]

In his book, *The Psychic Side of Sports*, co-authored with Rhea White and published in 1978, he documented many experiences of sportspeople and athletes that suggested the existence of telepathy between team members, out-of-the-body experiences, extraordinary feats of strength, and altered states of consciousness. He showed that sports and religion have many features in common:

In most religious teachings it is said that no lasting realisation can be achieved without many years of steady practice . . . Many athletes make that kind of commitment to their sport, at least for a part of their lives. The spiritually evocative elements we have discussed – long-term commitment, sustained concentration, creativity, self-integration, being in sacred times and place, and stretching to the limits of one's capacity – are common to both sport and religious discipline. Those similarities between the two kinds of activity often lead to the same kinds of experience.[31]

Murphy's magnum opus is *The Future of the Body: Explorations into the Further Evolution of Human Nature*,[32] published in 1992, in which he brings together evidence from more than 3,000 studies on exceptional human abilities in body, mind and spirit. Unlike most academic and scientific studies of human evolution that confine their attention to intellectual abilities and the developments of technology, Murphy includes psychic, spiritual and bodily abilities, giving a broad overview of human evolutionary potentials.

Murphy has influenced the thinking of many people within the sports world, and he has helped inspire a new organisation called the Sports, Energy and Consciousness Group that brings together star football players, extreme skiers, tennis players, rowers, coaches, scientists, psychologists, meditators and athletes. The mission of this group is 'to accelerate the global evolution of human consciousness by providing transformational practices that translate the wisdom of sport's "Ideal Performance State" into practical training methods that include energetics, awakened states of consciousness, and the unification of body, mind, and spirit.'[33]

From the 1970s onwards, another line of research converged with Murphy's approach: the study of positive psychology.

In the flow

Starting in the 1970s, research by positive psychologists showed that people's best moments are not when they are being passive, receptive or relaxed. Their most positive experiences usually occur when their body or mind is stretched to the limits in a voluntary effort to accomplish something difficult and worthwhile, summed up in the word 'flow'. To start with, these research studies looked at artists, athletes, musicians, chess masters and surgeons, because these were people who seemed to spend their time in activities that they enjoyed.

Their most optimal experiences depended on a sense of mastery, or of participation, in a state of flow.[34] People enter flow states through many different kinds of activity, including playing music and dancing, but sports are one of the most common ways in which people find themselves in the flow. Three conditions have to be met to achieve the flow state or 'being in the zone':

1. The activity should have a clear set of goals, giving the task structure and direction.
2. The task must have clear, immediate feedback, so that the performance can be adjusted to maintain the flow state.
3. There should be a good balance between the perceived challenges of the task at hand and the person's perceived skills. They must have a confidence in their ability to complete the task.[35]

Factors that inhibit the state of flow include apathy and boredom, which occur when the challenges are too low for a person's skill, and anxiety, which occurs when the challenges are too high.

Many traditional spiritual practices emphasise a need to be in the present, which can be achieved through meditation, for example, or through singing, chanting, and dancing (as I discuss in *SSP*, chapters 1 and 6). Sports provide an extraordinarily effective way of being fully present. One of their great advantages is that they provide clear goals and feedback. A tennis player knows what she has to do: to return the ball into the opponent's court and every time she hits the ball, she knows whether she has done well or not. A football team has a literally clear goal, namely to score more goals than the opposing team. In some other areas of human activity, like artistic creativity, the goals are less well defined or have to be defined by the artists themselves; the goals in sports are much clearer.

Sports also require a high degree of concentration. An avid rock climber, whose day job was as a professor of physics, said, 'It is as if my memory input has been cut off. All I can remember is the last thirty seconds, and all I can think ahead is the next five minutes.'[36] The racing driver Jochen Rindt observed that when driving, 'You completely ignore everything and just concentrate. You forget about the whole world and you just . . . are part of the car and the track . . . There is nothing like it.'[37]

A professional swimmer explained, 'When I've been happiest

with my performance, I've sort of felt one with the water and my stroke and everything.'[38] The professional golfer Tony Jacklin revealed, 'When I'm in this state, this cocoon of concentration, I'm living fully in the present, not moving out of it. I'm aware of every inch of my swing.'[39]

Participants in many other sports give similar descriptions of states of flow using a range of words to describe this state, including 'in the zone', 'focused', 'everything clicks', 'in the groove', 'tuned in', 'switched on', 'going really well', 'floating', and 'super alive'.[40]

Some sportspeople explicitly describe this state of flow as a spiritual experience. The basketball player Patsy Neal wrote, 'There are moments of glory that go beyond the human expectation, beyond the physical and emotional ability of the individual. Something *unexplainable* takes over and breathes life into the known life . . . call it a state of grace, or an act of faith . . . or an act of God. It is there, and the impossible becomes possible . . . the athlete goes beyond herself, she transcends the natural, she touches a piece of heaven and becomes the recipient of power from an unknown source.'[41]

The thrill of speed

One way of being in the flow in the most literal sense is through moving at high speed. Many people find speed exhilarating. Modern speed sports, like downhill skiing, motorcycle and car racing take human experience to levels that were unknown before the twentieth century, and aircraft take speed yet further. The airspeed records for pilots of jet planes are now over 2,000 miles per hour, about three times the speed of sound.

My own experience of excitement of speed began as a child in flying dreams, in which I could move at will very fast by flying through the air. Flying dreams are quite common, especially in children, and it may be that most people have their first experiences of speed in the dream world rather than in physical reality. In the course of human history, dreams of flying long preceded humans' technological ability to fly, and so did images of flying gods and goddesses, as in ancient Sumeria and Egypt, and in Jewish, Christian and Muslim images of angels as winged beings.

The dream of flying was also expressed in myths. In the ancient Greek story, Icarus acquired the ability to fly with wings made of feathers and wax, but he soared too close to the sun and the wax melted. He plunged to his doom in the sea. His *hubris*, his pride, was followed by his fall.

As a child, my first experience of the thrill of physical speed was riding downhill on a bicycle, which was far more exciting than running as fast as I could. Speed is not so exciting when it depends entirely on our own unaided muscular efforts, as in running or swimming. It takes on a special quality when the motive power is not our own, but we control it.

Before the advent of modern motorised technologies, there were several ways of experiencing speed that did not depend primarily on human muscular effort, like riding a galloping horse, kayaking down a fast-moving river, surfing ocean waves in Hawaii, skiing downhill in Norway and diving into the sea from a cliff. To these traditional pursuits, twentieth- and twenty-first-century technologies have added many new ways of speeding, including snowboards, in-line skates, motorbikes, cars, racing yachts, planes and skydiving.

Part of the exhilaration of speed comes from being in the flow, in the present, especially when small changes can have big effects. Subtle alterations in balance on skis, or minimal hand movements on the steering wheel of a racing car can make the difference between life and death. Indeed, danger is part of the thrill of speed. As the American writer Hunter S. Thompson wrote in relation to motorcycling, 'Faster, faster, faster, until the thrill of speed overcomes the fear of death.'[42] But even when that fear is overcome, the danger remains, even for professionals. Motorcycle racing has claimed hundreds of lives,[43] as has car racing.[44]

Ayrton Senna, one of the greatest racing drivers of all time, and three times Formula One champion, recalled that in 1988 he went to previously inaccessible levels of performance by his desire to beat a rival. When driving at enormous speed, he reached a mental state where his subconscious mind took over: 'Suddenly I realised that I was no longer driving the car consciously. I was kind of driving by instinct, only I was in a different dimension . . . It frightened me because I realised I was well beyond my conscious understanding.' This was like a religious experience for Senna,[45] who died in a racing accident in 1994.

We live in a world where health and safety legislation and liability litigation enforce strict standards of risk reduction in most of our everyday activities. But these concerns are literally thrown to the winds by those for whom the danger of death takes second place to the exhilaration of speed.

Group flow

In team games, players sometimes find themselves working together in ways that seem to go beyond luck or coincidence, or picking up subtle sensory clues. They seem to be telepathic with each other.

Jayne Torvill and Christopher Dean became the most famous ice-skating dancers of their generation, and were well-known for their extraordinary rapport. Their two bodies 'moved as one'. As Dean himself commented, 'We are telepathic on the ice. There's simply no other way to explain it.'[46] Sensory clues must have played a large part in their rapport, but there could well have been more to it, exactly as Dean said.

The famous Brazilian soccer player Pelé went further. It was said of him, 'Intuitively, at any instant, he seemed to know the position of all the other players on the field, and to see just what each man was going to do next.'[47] No doubt this was partly a matter of alertness, concentration and good peripheral vision, but more may be involved as well. Telepathy and the normal senses are not mutually exclusive, but may often work together.[48]

Teams are social groups, in which the individual members become like a single organism to achieve common goals, including the scoring of literal goals. The bonds between them can serve as channels for telepathic communication, as in other social groups. But not all individuals are bonded effectively, and not all teams function well as organisms. Even within a well-established team, whose members have had much shared experience, this state may come and go. It comes as a contagious confidence spreads through the team; it goes when the members are tired or demoralised. Michael Novak, a perceptive writer about sports, expressed it as follows:

> When a collection of individuals first jells as a team, truly begins to react as a five-headed or eleven-headed unit rather than as an aggregate of five or eleven individuals, you can almost hear the click: a new kind of reality comes into existence at a new level of human development . . . For those who have participated in a team that has known the click of communality, the experience is unforgettable, like that of having attained, for a while at least, a higher level of existence. [49]

Bonding between members of a group is of enormous importance in the armed forces, and military training programmes are designed to inculcate team spirit. Real combat experience is even more effective.

Similar experiences occur in many other kinds of group activities, including singing and the playing of music. Catherine Baker, a professional bassoonist, told me that musicians playing together use nonverbal communication all the time, sometimes including telepathic links between players, as well as between players and conductors. 'It does seem that when a chamber group orchestra gets this psychic "link" in a performance, the audience genuinely knows that it has been part of something special (as do the players!).'

Other musicians who play in folk music and jazz groups have told me much the same. So have actors who work in theatre groups, especially when they are improvising together. And so have people who dance together.

As I discuss in Chapter Two, telepathy in human groups is part of our evolutionary heritage and is rooted in the way that animal groups are coordinated through social fields, for instance when flocks of starlings change direction very rapidly as they fly without bumping into each other. Not only do the individual birds know where the other members of the flock are, but they also anticipate where they are going next. The same applies to schools of fish.[50]

The principles of group flow apply not only to teams themselves, but also to their supporters or fans, who are often linked together by their shared emotions, as well as by chants, songs and collective movements like Mexican waves. Players are often affected very positively by being bathed in an atmosphere of support. The sportscaster Jerry Remy noted that in home matches of the Boston Red Sox baseball team, the intensity of the fans' support 'not only boosts the Sox, but also intimidates the other players.'[51]

Negative effects of flow

Everything has a shadow side, and so does flow. People derived pleasure in ancient Rome from seeing gladiators fight to the death, or Christians devoured by lions, and many people still enjoy watching bullfights, cockfights and dogfights, as well as various forms of human fighting. The infliction of pain can be a source of pleasure for some people, as the Marquis de Sade made clear; his name, of course, gave us the word 'sadist'.

Some people find fighting in wars exhilarating. Veterans from Vietnam and other conflicts sometimes speak with nostalgia about front-line action, which they describe as a flow experience. One commented that when he was sitting in a trench next to a rocket launcher, life was focused very clearly. There was a clear goal, to destroy the enemy, and distractions were eliminated. Indeed, for many veterans, some of their war experiences were much more exhilarating than anything they encountered in civilian life.

Criminals often experience a state of flow when carrying out their crimes. One said, 'If you showed me something I can do that is as much fun as breaking into a house at night and stealing jewellery without waking anyone up, I would do it.' And many acts of juvenile delinquency like car theft, vandalism and rowdy behaviour are motivated by the desire for flow experiences.

Meanwhile, many video games are deliberately designed to encourage states of flow in those who play them,[52] applying the principles worked out by positive psychologists. This is one reason that they can be addictive: some people play for many hours a day, neglecting personal hygiene, losing regular sleep patterns and avoiding normal social life. The American Psychological Association has even come up with a definition for this problem: Internet Gaming Disorder.[53]

These are all examples of individual experiences of flow going in a destructive direction. It is also clearly the case that groups in a state of flow can be very dangerous, as in rioting crowds, violent sectarian conflicts, or lynch mobs.

However, none of these examples of the negative effects of flow experiences are arguments against the importance of flow itself, any more than sexual abuses are arguments against sex. The positive effect of flow and its spiritual value depend on the context, as I discuss below.

Oriental martial arts

Eastern martial arts, which developed principally in India, China, Korea and Japan, are more explicitly related to spiritual practices and traditions than in most other cultures. In particular, they explicitly recognise the importance of the flow of energy, which in India is called *prana* and in China and Japan *chi* or *qi* or *ki*. By contrast many Western athletes and sportspeople are taught by their coaches and trainers to think of their bodies as machines, following the standard scientific orthodoxy.

Dave Meggyesy, looking back on his time as a professional football player, reflected, 'I knew my body more thoroughly than most men are ever able to, but I had used it and thought of it as a machine, a thing that had to be well-oiled, well-fed, and well-taken-care-of, to do a specific job.'[54]

One of the first books to popularise the oriental approach in the West was Eugen Herrigel's *Zen in the Art of Archery: Training the Mind and Body to Become One*. Herrigel was a professor of philosophy who studied the Japanese art of archery, *kyudo,* when he was living in Japan in the 1920s. His teacher emphasised the spiritual aspect of this practice. One of the central ideas was that through years of practice, activity could become both mentally and physically effortless. Mastery involves the development of habits that no longer require conscious control.

Herrigel's book, first published in German in 1948 and in English in 1953, helped inspire a widespread interest in Zen Buddhism, and also an approach to sports that emphasised emotional, mental and spiritual elements, as well as purely physical ones. He described in detail his lessons over a period of years, and his master's emphasis on his state of mind, not simply on physical abilities:

I once remarked that I was conscientiously making an effort to keep relaxed. He replied: 'That's just the trouble, you make an effort to think about it. Concentrate entirely on your breathing, as if you had nothing else to do!' It took me a considerable time before I succeeded in doing what the master wanted. But I succeeded. I learned to lose myself so effortlessly in the breathing that I sometimes had the feeling that I myself was not breathing, but strange as this may sound, being breathed.[55]

His studies combined a regular routine with a spiritual dimension. As the master taught him, 'It is necessary for the archer to become, in spite of himself, an unmoved centre. Then comes the supreme and ultimate miracle: art becomes "artless", shooting becomes not-shooting, a shooting without bow and arrow; the teacher becomes a pupil again, the master a beginner, the end a beginning, and the beginning perfection.'

Some schools of martial arts emphasise that the true opponent to be overcome is oneself, one's fears and frustrations and self-limiting concepts. In karate training, the student may go through the repetition of a formal exercise over and over again, beyond the point of exhaustion. He may then reach a point where he simply no longer has the strength to maintain his tense habits of contracting body and mind, and his movement takes on a naturalness and fluidity that he had not suspected possible.[56]

The skill consists in directing the energy, or *chi,* through intention, and the intention then directs the body. The aim is not to use the body directly, with brute muscular force, but to use the will to direct the energy, forming a sensitive feeling of movement; then the physical movement follows.[57] Some masters of aikido are said to have developed the capacity to throw opponents without touching them, or knock them down at a distance.[58] At a high level of mastery, the techniques involve a direct influence on an opponent's mind. One of the skills that advanced practitioners develop is the ability intuitively to anticipate the opponent's next movement.

In terms of self-defence, part of the art is not to incite aggression in the first place, in other words to have an attitude that creates an atmosphere in which the potential aggressor does not have an aggressive impulse.

A Zen story emphasises this point.

A sword master wanted to test three of his students and took them to a narrow ravine in which lived a fierce wild horse. He told them to come through the ravine one at a time while he waited at the other end. The first student started on his way and when the wild horse lunged at him, he skilfully blocked and dodged it, and won his way through to where the master was waiting. He then turned to see if his fellow students would put up as good a fight.

The next student saw the horse from a distance and nimbly climbed up the walls of the ravine, passing far above the horse's head, well out of reach. He joined his friend next to the master. Both turned to watch the performance of the last student. Would he avoid the animal or engage it in fierce combat? He appeared at the mouth of the ravine, looking calm and unconcerned. As he walked through the ravine, the wild horse whinnied in greeting and simply let him pass.

When I was a graduate student in Cambridge, I myself learned aikido for about a year. I claim no special skill or distinction, and I only put my training into practice on one occasion. Some twelve years after my brief period of study, I was at a conference in the United States and after going out for dinner with some friends, I returned to the place I was staying. In the car park I saw one of the other speakers, an international celebrity, beating up a woman. I told him to stop it. He then turned on me and came towards me, obviously drunk, with his fists flailing.

Without thinking about it, I suddenly became surprisingly calm and my energy centred in the region of my *hara*, near the navel, in accordance with my aikido training. I stood there waiting for him, trusting I would know what to do. When he came within a few feet of me, he collapsed onto the ground!

Extreme sports

In the twentieth and twenty-first centuries there has been an extraordinary growth of extreme sports, including climbing without ropes – so-called freeclimbing – on mountains and skyscrapers, skydiving, freediving without breathing equipment, extreme marathons, and BASE jumping, leaping off a fixed structure in a wingsuit, or with a parachute. The acronym BASE refers to the kinds of structure from which the jumpers leap, Buildings, Antennas, Spans (like bridges) and Earth, like cliffs.

The American alpinist Dean Potter climbed almost impossibly difficult places in the mountains of Patagonia, solo and without any form of protection; he free-climbed the Half Dome and El Capitan in Yosemite National Park in a single day, and he also took up BASE jumping in his thirties, throwing himself off bridges, buildings and cliffs, opening his parachute at the last possible second.

When asked why he did all these things, he said it was not purely for sport or adventure, but for enlightenment. These activities were his spiritual path: 'When I'm jumping, climbing or highlining [walking on a kind of tightrope made of flat webbing], it's as truthful as I can be. What I'm seeking when I do things is to break free from all my attachments.'[59] He took risks that could have killed him, and he said that the fear this induces was the only thing that brought out his full potential.

Other extreme sports depend on pain rather than immediate physical danger. Ultramarathons include the annual Trans-American Footrace, which is more than 3,000 miles long, and run in sixty-four consecutive daily stages.

One of the cycling equivalents is the race across America,

RAAM, 3,042 miles from the west to the east coast of the US, probably the longest and toughest cycle event in the world. There are no designated rest periods, and the clock keeps running from start to finish. To win, a cyclist must be prepared to ride for twenty-two hours a day over mountain passes and through deserts, in chilling rainstorms and in blistering heat. The lack of sleep, the constant effort and the stress of competition take their toll in many ways.

A similar cultivation of punishing pain underlies participants in the 135-mile non-stop Badwater Ultramarathon running race across Death Valley in California, which is held in July, when daytime temperatures can rise to 129 degrees Fahrenheit or 54 degrees Celsius. One of the participants, Kirk Johnson, called this experience a 'portal' into another realm of experience:

> If there was a place where human limitation, but also the limits of explanation and reason and science, should hit the wall, this was it. Badwater was a perch from which I could look for the definitions of what we are, what makes us stop and what makes us go. Maybe spirituality wasn't quite the word for what I thought, but I was enough of a believer, or a seeker at least, to think that there might be a way . . . to reach the veil and touch something beyond me in my life. A place where misery and transcendence were so deeply intertwined, it could not be without meaning.[60]

In the 1970s, researchers coined the term 'risk exercise response' for the exhilaration and euphoria that people experience during high-risk sports. These experiences are associated with a cascade of chemical changes in the brain, including elevated levels of the neurotransmitter dopamine, associated with reward circuits in the brain; endorphins, which are natural morphine-like compounds; and the neurotransmitter serotonin. Added to this cocktail are adrenaline and testosterone in the bloodstream.[61]

Drugs that increase the amount of dopamine in the brain, such as cocaine and methamphetamine, seem to speed up the internal clock so that time appears to go more slowly. So do moments of high stress, when dopamine floods the brain.[62]

Freediving is one of the most dangerous of extreme sports. The divers go to great depths, sometimes more than 300 feet, while holding their breath and without the use of any diving equipment. Of the 10,000 active freedivers in the United States, about twenty die every year, which works out to about 1 in 500. This makes it the second most dangerous adventure sport, after BASE jumping, where the death rate is around 1 in 60.[63]

But freediving is not new. Pearl divers flourished in the South Pacific, Persian Gulf and Asia for more than 3,000 years. When Marco Polo visited Ceylon in the fourteenth century, he saw pearl divers plummeting more than 120 feet. Around 700 BC, Homer wrote about divers who lashed themselves to heavy rocks and plunged more than 100 feet deep to cut sponges from the seafloor.

Right up into the twenty-first century, female Japanese sea divers called *ama*, meaning 'sea women', harvested food from the seafloor. Very few still do this, but in the nineteenth century, thousands worked off the east coast of Japan, and European sailors reported that they plunged to depths of hundreds of feet on a single breath, staying underwater for up to fifteen minutes at a time.[64] In 2009, the world record for holding breath underwater under rigorously monitored conditions was 11 minutes and 39 seconds.[65]

People training to become freedivers practise holding their breath underwater in swimming pools, and get used to the bodily changes that occur. Research in the late twentieth century showed that when people are submerged and holding their breath, a physio-logical 'master switch' leads to a dramatic reduction in their heart rate and a shift of blood circulation away from the limbs and towards the vital internal organs. Similar changes occur in seals and other marine mammals, enabling organs like the brain and the heart to remain oxygenated longer, and extending the amount of time they can stay submerged.[66]

Many freedivers talk in quasi-religious terms about the 'doorway to the deep' as transcendent, life changing, or purifying. They find themselves in 'a new and shimmering universe.'[67] Some describe it as 'stillness' or as 'full-body meditation'.[68] But when freediving becomes a competitive sport, it is all about numbers, about beating records or competitors.

The journalist James Nestor, who wrote about freediving in his book *Deep*, went to several competitive dives: 'Most of the competitive divers I met seemed to have little interest in exploring the deep ocean that they had painstakingly trained their bodies to enter. They dived with their eyes closed; nitrogen narcosis struck them dumb; they forgot where they were and why they were there. The deepest divers lulled themselves into a catatonic state that removed *any* sense of actually being in the water. The aim: hitting a number on a rope. Beating your opponents. Winning a medal.'[69]

This obsession can be lethal. In 2012, Nicholas Mevoli, a thirty-two-year-old from Brooklyn, attempted a 314-foot dive to set a new record. He did not make it and had to be dragged back to the surface, with blood dripping from his mouth. As he himself wrote before this incident, 'Numbers infected my head like a virus and the need to achieve became an obsession.' The following day he tried to break another record, recklessly going beyond his limits. He was still conscious when he reached the surface, but blood was pouring from his mouth and his pulse gave out after fifteen minutes. All attempts to resuscitate him failed.[70]

Some competitive freedivers see setting records not simply as a personal goal, but also as a process of extending human potential.[71]

The evolution of human potential through sport

As Michael Murphy pointed out, sports are the principal way in which human physical potentials are evolving in the modern world. They are doing so in several ways. First, through the continual breaking of records and improvements of performances. In 1954, Roger Bannister first ran a mile in under four minutes, and this is now a standard achievement for professional runners; the record is now 3 minutes 43 seconds. Many other sports records are continually broken, both in athletic events and also in sports that rely on technologies, like motorcycle racing and skydiving.

Second, sports are evolving through a natural selection of techniques in mixed martial arts (MMA), in which practitioners of different martial arts are pitted against each other. Such contests occur under the aegis of the Ultimate Fighting Championship or UFC, and in principle allow for the natural selection of different fighting methods and the evolution of new combinations of martial arts techniques. At the first Ultimate Fighting Championship in 1993, there were virtually no rules, only a prohibition against biting and eye-gouging, and there were also no referee stoppages. As these fights proceeded, there were scenes of increasing brutality, and in the eighth UFC the winner trapped his hapless opponent on the ground and proceeded to smash in his face with eight devastating elbow blows.[72]

By 1996 there were widespread protests against MMA, which the US Senator John McCain described as 'human cockfighting'. In order to prevent a ban, new rules were introduced to make the sport more acceptable. So far, the result of this evolutionary process has been the predominance of boxing, kickboxing, wrestling, judo and ju-jitsu. These are the fighting techniques that have proved most effective when confronted with other martial arts.[73]

UFC has also evolved to become heavily commercial, with armies of trainers, managers, and sports psychologists.

Third, entirely new skills and experiences are continually being developed, like snowboarding, windsurfing, hang-gliding and in-line skating. Each of these sports involves not only the evolution of skills and abilities, but also of the technologies that support them, with ever-more sophisticated designs and construction materials. Humans can now do things that their ancestors could only have dreamed of, such as skydiving, plummeting through the air from a height of 12,000 feet.

The range of human experiences is increasing dramatically, and most of this evolution is not impelled by utilitarian concerns. Nor is it taking place primarily for military reasons, or to make money. It is an expression of an exploratory urge, a creative impulse. Some of the competitive aspects of this sport may well be driven by motives that Charles Darwin would have recognised in his theory of sexual selection. Triumphant figures in these sports may well be more attractive to potential mates. But sexual competition alone cannot explain this gratuitous evolutionary creativity.

Finally, a greater understanding of the psychology of sports is itself helping to drive the evolutionary process. In highly competitive events, very small improvements in performance can give people an edge over their competitors, and some sports training programmes now include meditation, 'inner-game' visualisation techniques, and the deliberate cultivation of flow experiences, because these can lead to the enhancement of performance.[74]

The flow of the spirit

How can we understand the spiritual experiences that many people have when engaged in sports? Part of this effect depends on being in the present, rather than being taken out of it by worries, anxieties, regrets about the past and other kinds of rumination. Meditation can help in the achievement of a state of presence, but sports often do so quicker and more effectively. But they do so in a different way.

The spiritual experiences that occur during meditation are often described as being beyond time and space. They are not so much an awareness of change as an awareness of a timeless ground of being. By contrast, in sports, people experience being in the flow, literally in a process of movement. This corresponds to a different aspect of spiritual reality. As I discuss in Chapter Eight, in many traditions there are three-fold models of spiritual reality: a ground of being; a principle of form; and a principle of energy and bliss.

One Hindu version of this trinity is called *sat-chit-ananda*, being-consciousness-bliss. In the Christian Holy Trinity, the Father is the ground of being; the Son or *Logos* the source of form; and the Holy Spirit the principle of flow and energy. Whereas meditation can lead to an experience of connection with the ground of being, or *sat*, sports are more related to experiences of flow or spirit, or *ananda*, which are inherently blissful.

At the physiological level, many aspects of sports that enable people to experience states of bliss or self-transcendence are associated with increased levels of hormones, such as adrenaline and testosterone, and with increases in levels of neurotransmitters in the brain, such as dopamine. Similar changes occur in many other animal species, including rats, when they are engaged in challenging physical activities.[75]

This does not mean that the spiritual side of sports is simply biological, as opposed to spiritual. If non-human animals show similar physiological changes to humans when they are engaged in physical activities, they may also have spiritual experiences.

Do hawks soaring in the sky experience the thrill of speed? Do dolphins leaping through the bow waves of boats experience joy in their freedom of movement? Do predators running at high speed while chasing their prey animals experience a sense of presence in the flow? Are they 'in the zone'?

I think that the answer to these questions is probably 'yes'. Non-human animals may well have experiences of being in the flow. In fact, some animals may have them more often and more intensely than humans, because they are not subject to the same distractions of thought, worry and egotism. And many animals seem to play for the joy of playing.[76]

Some people may disagree, and seek to confine spiritual experiences to human beings. But if a more-than-human consciousness underlies not just human nature but all nature, then spiritual experiences may be very widespread in the natural world. Many species of animals may be able to participate in the joy or bliss of the spiritual realm. Humans may differ from many other species of animals in their ability *not* to be in the flow, by being preoccupied with worries, ruminations, and self-centred fantasies. As practitioners of martial arts realised long ago, the physical practices of sport can help humans achieve a unity of mind and body that comes naturally to non-human animals, but from which we are often alienated.

Yet only humans can integrate these experiences with conceptions of nature, mind and reality. I continue this discussion in Chapter Eight.

Two ways of experiencing flow through sports

HONE YOUR SKILLS

Being in the flow through sports can only happen if you have a sufficient level of skill. You need to feel confident in what you are doing and also to feel stretched by the activity in which you are participating. Many people are already practising these skills and experiencing states of flow, but if you are not already doing so, and if you want to experience this way of being in the present, you will have to practise a sport you already know or take up a new one.

It need not require a highly strenuous activity, but must involve a level of skill. Ping-pong, for example, can facilitate a sense of flow just as tennis can, although the activity level is less. For the sheer enjoyment of physical skill, it is necessary to acquire the physical skill first.

PARTICIPATE BY SPECTATING

When watching a spectator sport on television, like football or tennis matches or horse races, become aware of the engagement of the sportspeople in that flow, and also the participation of the spectators. Best of all, go to a live event and become aware of those moments when the players seem to be in the flow and the spectators are fully engaged with them.

You can feel this engagement yourself and become part of it, experiencing a collective flow and its ups and downs in disappointment and exhilaration.

2

Learning from Animals

Connecting with animals

When I was a child we had a dog, a Scots terrier called Scamp, and I kept many other pets including pigeons, a jackdaw, a budgerigar, a rabbit, hamsters, white mice and two tortoises. For several years I collected frogspawn and raised tadpoles, and watched them turn into little frogs. I kept peacock butterfly caterpillars that I fed on nettles, and saw them pupate and emerge as gloriously coloured butterflies, which I then released; I did the same with cinnabar moth caterpillars that I fed on ragwort. These transformations amazed and intrigued me. But when I went away to boarding school I was not allowed pets, and the only animals I could keep were fruit flies, which I bred for genetics experiments.

One of my favourite books was Jean Henri Fabre's *Book of Insects*. His chapters on the life histories of insects were the result of many patient hours of observation in the wild, and, among other wonders, he described the musical talents of the locust and the sexual cannibalism of the praying mantis, where having sex is very risky for males, which are often eaten by their mates. Charles Darwin described Fabre as 'an incomparable observer'.[1]

I was fortunate that my father was interested in natural history, and encouraged my curiosity. I was like many other people, especially children, who are curious about non-human animals. Millions of people visit zoos, often as families. Natural history films are immensely popular on television, and animal videos often go viral on the internet. Even in cities, people enjoy watching wild birds, feeding them, and listening to their songs. In Britain, the largest conservation charity is the Royal Society for the Protection of Birds, with more than a million members.[2] But our most direct relationships are with pets, as I discuss below.

Animals can connect us to the realm of the spirit through their beauty, grace and elegance, through their freedom of movement, and through their many powers and abilities. Some of our abilities far exceed those of other species, but many non-human animals have abilities that surpass our own, such as the flying of birds, the swimming of fish, the leaping of monkeys from tree to tree, the complex societies of millions of termites who build huge architectural structures even though they are blind, the migrating of Monarch butterflies over thousands of miles, and the echo-location system of bats.

Tardigrades, also known as water bears, are tiny animals that live in moss, 0.5 mm long when fully grown. They far exceed our own powers of survival. They can tolerate extreme cold, a few degrees above absolute zero; and withstand temperatures of 150 degrees Celsius; they can survive pressures greater than those in the deepest ocean trenches, and also withstand the vacuum of outer space; they can undergo repeated dehydration and rehydration, and stay alive without food or water for more than thirty years.

The very different abilities of animals remind us of our own limitations. One of the most vivid investigations of these differences was by a British lawyer and veterinarian, Charles Foster, which he summarised in his book, *Being A Beast*. He was inspired by a famous question asked by the philosopher Thomas Nagel: 'What is it like to be a bat?'

Foster wanted to know what it was like to be a badger, an otter, a deer, a fox and a swift, so he tried to follow their ways of life. He spent weeks living in a hole in the ground like a badger, with his eight-year-old son, eating earthworms. Being an otter, he swam in the rivers of North Devon at night. He scavenged from the dustbins in London's East End like an urban fox and followed a swift's migration route from England to West Africa (on the ground). Here is a short extract from his account of life as a badger:

> Quite a lot of being a badger consisted simply in allowing the wood to do to us what it did to a badger; being there when it rained; keeping badgers' hours; being cramped underground (there's no possibility of thinking the world is at your sovereign feet when in fact it's over your head, squashing your legs and dropping into your eyes); letting the bluebells brush your face instead of your boots. But there were some high physiological fences keeping us out of the badger's world. The main one is scent . . . A badger's landscape is primarily a scent landscape.

Foster described in detail the changing smellscape at badger level throughout the day, in rainstorms and in winter but, of course, however hard he tried, he could not match the badger's experience or sensitivity to smells.

There is a vast body of scientific research about animals, and it tells us a lot about these other creatures with whom we share our world. But it does not tell us what it's like to be another kind of animal. Charles Foster tried to find out, but inevitably came up against the limitations of human senses and abilities.

In this chapter, I first discuss the history of human relationships with animals, and consider some of their powers and abilities that exceed our own. I then look at animal telepathy, which illuminates some of our own unexplained abilities. I end by considering how learning from animals can help us on our spiritual journeys.

The evolution of human-animal relationships

Our archaic human ancestors, collectively known as hominins, lived as hunters and gatherers for more than six million years[3] before the emergence of our own species, *Homo sapiens*, some 100,000 years ago.[4] Today, in the few surviving hunter-gatherer societies, only a small proportion of their food comes from hunting; most comes from gathering. The exceptions are the hunter-gatherers of the plant-poor Arctic.[5] Hominins and early *Homo sapiens* may also have relied primarily on gathering, and probably obtained more meat by scavenging from kills left by more effective predators such as big cats than by hunting.[6] Big-game hunting, as opposed to scavenging, may date back only some 70,000 to 90,000 years.

Hunter-gatherers need to know what plants are edible, which have healing powers, where they grow, when they fruit, and how to make food from them. They also need to find the animals they hunt, and to know their habits. In hunter-gatherer cultures, human beings do not generally think of themselves as separate from the realm of other animals, but as intimately interconnected.[7] The world around them is alive. They are animists. Their mythologies portray the interconnectedness of human and non-human forms of life, and often see them as sharing a common origin.[8]

In hunter-gatherer societies, some people specialised in connecting the human and non-human worlds. Such intermediaries are generally known as shamans, using a term borrowed from tribes in Siberia. They formed a relationship with the wild animals among which they lived, and in trances and dreams communicated with animal spirits, experiencing themselves as being guided by animals, understanding their language, and sharing in their powers.[9]

The first animals to be domesticated were dogs, long before any other species. Dogs' ancestors, wolves, hunted in packs.

Peoples also hunted in groups, and from an early stage dogs were used for hunting as well as for guarding human settlements. Their domestication predated the development of agriculture,[10] and they were the only animals to be domesticated before people adopted a settled way of life.[11] Some evidence from the study of DNA in dogs and wolves points to a date for the transformation of wolves to dogs over 100,000 years ago. This DNA evidence also suggests that wolves were domesticated several times, not just once, and that dogs have continued to interbreed with wolves.[12]

Our ancient companionship with dogs may have played an important part in our own evolution; and dogs may have had an essential role in the domestication of other species, both through their ability to herd animals such as sheep, and also by helping to protect flocks against predators.

In 2009, an international team of scientists announced that they had identified the earliest archaeological evidence of a dog in the Goyet caves in Belgium, dating back 31,700 years. It probably resembled a Siberian husky, but was somewhat larger, and subsisted on a diet of horse, musk ox and reindeer meat. Other early evidence was found in the deepest part of the Chauvet cave in France with a track of footprints of a large dog walking with a child. Soot on the roof of the cave dates from 26,000 years ago, and probably came from flaming torches.

The wolves that became dogs have been enormously successful in evolutionary terms. They are to be found everywhere in the inhabited world, hundreds of millions of them. The descendants of the wolves that remained wolves are sparsely distributed, and often endangered.

By the time of ancient Egypt, there were already several distinct breeds: a greyhound or saluki type, a mastiff type, a basenji type, a pointer type, and a small terrier-like type.[13] Dogs were venerated and some were embalmed; in every town in ancient Egypt a graveyard was devoted entirely to dog burials. The god of the dead was the dog- or jackal-headed Anubis.

In today's world, there are great cultural variations in the way dogs are treated. In Arab countries and in the Indian subcontinent, they are generally shunned, and there are large populations of stray or feral dogs, a source of dangerous diseases such as rabies. Even so, individual hunting dogs are admired and pampered. In other parts of the world – China, Burma, Indonesia and Polynesia – dogs are slaughtered for human food.[14] But in many cultures they are generally treated affectionately.[15]

The domestication of dogs happened so long ago that we will never know the details, but a twentieth-century study in Russia with silver foxes showed that quite rapid changes occur under conditions of selective breeding. From the 1950s onwards, generations of silver foxes were bred in captivity, and the tamest foxes were selected as the parents of the next generation. After forty generations, the Russians succeeded in producing a breed of silver foxes that are docile, friendly and as skilled as dogs in communicating with people.[16] These tame foxes also look different from their wild ancestors, with broader heads and juvenile features.[17] Some are now being sold as pets.

Francis Galton, Charles Darwin's cousin, pointed out that relatively few species were suitable for domestication. Animals capable of being domesticated had to be hardy, and survive with little care and attention. They also had to be gregarious, and hence easy to control in groups. Sheep, goats, cattle, horses, pigs, hens, ducks and geese all met these criteria. But other gregarious species, such as deer and zebra, despite many attempts at domestication, still remain too wild to manage with ease.[18]

Cats are the only domesticated animals that are not gregarious, but through their territorial and comfort-loving natures they form symbiotic relationships with people, while retaining something of their independence as solitary hunters. They revert with relative ease to a free-living, feral existence.[19] Cats were domesticated much more recently than dogs, probably no more than 10,000 years ago. The oldest archaeological evidence for domestic cats comes from Crete, about 9,500 years ago, and cat remains from Jericho have been dated to 8,700 years ago.[20]

The first historical records of cats are from ancient Egypt, where they were treated as sacred, and killing them was forbidden. Some 3,600 years ago, house cats were depicted in Egyptian tomb paintings; they were also mummified in enormous numbers. At the beginning of the twentieth century, with the growth of international trade, cat mummies were excavated by the ton, ground up and exported as fertiliser to England.[21]

Horses were also domesticated relatively recently, probably only about 5,000 years ago in the region around Turkestan. They may have been first used as draught animals, but they soon became important in war and in hunting, when they were more like comrades than slaves.

In early civilisations, there was still a pervasive sense of human-animal connectedness. Many kinds of animals were regarded as sacred, and in modern India, cows, elephants and monkeys are still viewed as sacred by many millions of Hindus. Many of the Hindu gods and goddesses took animal forms or had animal helpers, as they do to this day.

At first sight, there is little trace of this sense of solidarity with the animal kingdom in industrial societies. Machines have replaced beasts of burden. Horses, donkeys, mules and bullocks are no longer our daily companions. The intimate familiarity with traditional farm animals has been replaced by modern agribusiness, with animals kept in factory farms and industrial-scale feed lots.

Nevertheless, in our private lives, the ancient affinity with other animals remains. There are countless bird-watchers, naturalists and wildlife photographers. But these bonds are maintained most intimately through the keeping of pets. Even though most people in modern cities no longer need cats for catching mice, or dogs for herding or hunting, these animals are still kept in their millions, together with a host of other creatures that play no utilitarian role: ponies, parrots, budgerigars, canaries, rabbits, guinea pigs, gerbils, hamsters, goldfish, lizards, and stick insects.

Most of us seem to need animals as part of our lives. Human nature is bound up with animal nature. We may be diminished when we are isolated from other animals.

Pets

In ancient Egypt, as well as the larger dogs used for hunting, guarding and herding, smaller breeds lived in houses as pets. Ancient Greeks and Romans also kept house dogs. In Tibet and China, it was customary to keep both guard dogs and house dogs. Guard dogs were big and fierce and lived outside, while the small dogs lived indoors.[22]

Pet-keeping, as opposed to the keeping of animals for utilitarian reasons, was a luxury in ancient times. Now many more people can afford to keep pets. In some industrialised countries like France and the United States, most households contain at least one companion animal. Over recent decades, as urbanisation and prosperity have increased, more rather than fewer households have kept pets.[23]

The animal-keeping habits of different nations probably play a large part in the forming of national character. But there has been almost no systematic research; we have only bare statistics. Some of the highest levels of pet ownership are in the US: in 2012, thirty-seven per cent of households had dogs and thirty per cent cats, with a total of more than 43 million dogs and 36 million cats.[24] In the UK, in 2015, twenty-four per cent of households had dogs and seventeen per cent cats.[25] One of the lowest levels of dog ownership is in Germany where, in 2013, only fourteen per cent of households had dogs; nineteen per cent had cats.[26]

In most countries, there are more dogs than cats, but in Germany, Switzerland and Austria there are more cats than dogs. Many households keep birds, reptiles, fish, and small mammals such as rabbits, guinea pigs, hamsters and ferrets. In surveys about attitudes to pets, sixty-three per cent of pet owners said they considered their animals to be family members. Only one per cent thought of them as property.

In addition to pet dogs, there are many that help people in very practical ways, including sheepdogs and other working dogs, and service dogs play a vital role in the lives of many thousands of people. The best known are guide dogs for blind people, but there are also hearing dogs for deaf people, dogs that assist disabled people, and dogs that alert epileptics to oncoming seizures. Some pets are even officially classified as 'emotional-support animals'.[27]

There are also many programmes – more than two thousand in the United States – in which animals visit people in hospitals, hospices and homes for the elderly. These animals usually belong to volunteers and are often called PAT (Pet As Therapy) animals. They are helpful for children, especially for the chronically sick, many of whom eagerly await their animal visitors.[28] They are also popular among elderly people and among people in hospices, where they can have a relaxing effect on dying patients and staff, lightening the mood, and providing affection and physical contact.[29]

Over and above all these practical roles that animals play, forming relationships with animals is important for hundreds of millions of urban people, even though it is no longer economically necessary. Why?

First of all, it is deeply habitual for humans to be connected to non-human animals. And when we relate to them, we inevitably recognise that they have much in common with us. They share the same kinds of senses: taste, smell, sight, hearing and touch. Indeed their sight, hearing and sense of smell are often better than ours. Like us, they feel hungry and thirsty and need to eat, drink, and excrete; they have similar emotions, like fear, affection, anger, joy, sexual desire and sadness. We feel an affinity with them despite their difference from us. We can feel affection for them, and empathy, and love.

Above all, love is what binds most people to their dogs and cats. This love flows both ways. As Charles Darwin wrote in *On the Origin of Species* in 1859, 'It is scarcely possible to doubt that the love of man has become instinctive in the dog.'[30] In his book, *Dogs Never Lie About Love*, the psychologist Jeffrey Masson calls love the very *raison d'être* for a dog.[31] Many dogs outshine humans in their loyalty, and they far exceed most people's capacity for unconditional love.

Just as we grieve beloved people when they die, so we grieve beloved dogs, cats, horses and other animals. One of the things that pet animals can give us, and give our children, is the direct experience that death is part of life. They have shorter lives than we do. In the modern world, for many people the death of a pet is the first experience of the death of a loved one.

There is no doubt that some people love their companion animals, and that some companion animals love their people, and often this love is mutual. But what is love?

For materialists, love is a genetically programmed instinct, a form of behaviour programmed by selfish genes to ensure their own survival. The experience of love depends on hormones like oxytocin, on reward circuits in the brain, and on chemical neurotransmitters. Animals have much the same brain chemistry as we do and have similar emotional-activation circuitry in their brains.

Both human and non-human animals have limbic systems within their brains that are closely involved with the regulation of emotions, especially in the amygdala. The limbic system was present in the ancestors of reptiles, mammals and in the birds. It is an ancient emotional-activation system that we share with countless other species, and, in particular, with our beloved pets. The love we feel is evolutionarily ancient.

For people with a holistic worldview, these mechanistic explanations are fine as far as they go, but they tend to ignore the fact that bonds of love come from our abilities to relate to a person, or animal, or family, or larger group greater than ourselves. They are mutual bonds of exchange and relationship. There is a two-way flow. Individuals are interconnected. Love is not solely a matter of genetic programmes, brain circuits, electrical impulses, neurotransmitters and hormones inside individuals; love comes through relationships with others.

For religious people, love is a spiritual quality as well as a social and a biological instinct. One of God's principal aspects is love, according to Hinduism, Judaism, Christianity, Islam and other religions. When we show animals love, and they show us love, we participate in God's nature.

Animals that comfort and heal

Companion animals not only keep us company and give us joy, but they also help us when ill or depressed. My wife Jill named our cat Remedy for a good reason. Remedy sensed when she was needed, and sat or lay on Jill or me, working her healing magic through her warmth and her purr.

In the course of my research on the relationships between human and non-human animals over the last twenty-five years, I have built up an extensive database of more than 5,000 stories, including more than seven hundred accounts of animals that comfort and heal. Most of them are about cats and dogs that stayed close to people when they were sick or sorrowful.

There were many comments about dogs like these: 'My dog senses exactly when I do not feel well or am sad.' 'When I am sad, she does not leave me and puts her head on my knees.' One of the simplest yet most eloquent was from a woman in St Helens, Lancashire: 'I am autistic and have a dog Nickita, she knows how I am. She comforts me before I have told her. Sometimes I have bad days. She is there with me where I am.'

Many dogs seem to know when their person is ill and they stay close and behave in a truly comforting way. Some dogs even seem to know what part of the person's body is painful. A retired policeman in Poole, Dorset, often took out his daughter's collie Ben for walks. He believed dogs should not be allowed onto beds, but on one occasion when his back was very painful, he had to lie down: 'As my head hit the pillow, the bedroom door opened and in came Ben. He jumped up onto the bed and stretched out against my back. I felt too ill to say anything, but the feel of him against my back was good. He must have sensed that I was unwell and needed warmth.'

The responsiveness of cats is especially striking in animals that normally cherish their independence. A woman in Leoben, Austria, told me that her male cat Baerli 'loved his freedom. When I did not feel well or was sad, though, he never parted with me. Instead he lay on my lap purring and pressing himself closely to me. When I was well again, he was off as usual, especially at night.'

Pets also help people who are bereaved. Several studies have shown that after the loss of a spouse or partner, pet owners were less depressed and less prone to feelings of despair and isolation than those without pets; they also had better general health and needed less medication.[32] There are many such stories on my database. A woman in Paris told me, 'Both my cats stuck with me as if they didn't want to leave me alone with my sorrow, and this lasted exactly the time I was mourning. After that the cats were more aloof again.'

Mutual help is an essential aspect of social life in many animal species. Even scientists who believe that animal behaviour is shaped by selfish genes[33] acknowledge the importance of altruistic behaviour in ant colonies, in parental care in birds and mammals, and in social groups of every kind.[34] For instance, when a member of a herd or flock gives an alarm signal, alerting other members of the group to danger, it may be endangering itself by attracting the attention of a predator.[35]

The ability of animals to comfort and heal is not a sentimental fantasy, or an anthropomorphic projection, or merely anecdotal. Their beneficial effects have been quantified in many scientific studies.[36] A review of research funded by the US National Institutes of Health (NIH) showed a wide range of benefits of pet ownership on psychological and physical health in adults and children.[37]

In one study, elderly people in Philadelphia who adopted cats were compared with a similar group of elderly people who did not adopt cats. Regular follow-up interviews and tests showed striking differences between the two groups within a year. The cat owners felt better, while the non-owners felt worse. Measured by standard psychological tests, those with cats felt less lonely, less anxious and less depressed.

Cats also had a favourable effect in lowering blood pressure in people with hypertension, reducing the need for medication.[38] The benefits they conferred depended on the bonds between the people and the cats. Bonded cats provided fun, company and affection; they helped take people's minds off their troubles and their ailments. The stronger the bond, the greater the positive effects.[39] Likewise, relationships with dogs have many physiological benefits including the reduction of blood pressure.[40] The dogs themselves experience similar benefits, and their heart rates drop while they are being petted.[41] In a study of pet owners after they had been hospitalised with heart disease, those with pets showed an improved survival rate a year later compared to a control group of non-pet owners.[42]

For children, pets can offer both companionship and security, responding to demands and giving uncritical sympathy.[43] Disturbed children seem to rely especially heavily on animals as a source of support. In one study, delinquent adolescents were found to be twice as likely to talk to their pets and three times as likely to seek out their animal's company when lonely or bored.[44] Pets can also help children to develop a better sense of mutuality and responsibility as they look after them.[45] They can help to smooth relationships and improve family dynamics. In a study in Baltimore, Maryland, on the impact of pets on family life, many families felt closer, spent more time playing together and argued less after they adopted their pets.[46]

Prisons that allow animals to visit prisoners or allow prisoners to keep pets have seen a reduction in violence, suicides and drug use, as well as improved relations between prisoners and staff.[47]

Although most studies have reinforced the message that pets are good for you, this is not always the case.[48] Some people acquire a pet because they want benefits from it, and then abandon the pet if it develops behavioural problems or fails to make them feel better. But these unfortunate experiences are exceptions.

How is it that animals can be so beneficial to humans? How do they have healing effects? Attempts to explain their influence include words like 'empathy', 'acceptance', 'companionship', 'emotional security' and 'affection'. These are the same words that are often applied to the healing effects of other people. The secret of this healing power seems to be the same, whether it comes from people or from non-human animals: acceptance and love.

Critics might argue that the health benefits of feeling loved by humans or by companion animals are nothing but placebo effects. However, to call them placebos does not explain them but only gives them another name. Why do placebos work? The fact is that blank pills given to patients in double-blind clinical trials often make them better, because they have positive expectations. The expectation of getting better helps those people to get better.[49] Conversely, people who are lonely, depressed and without hope are likely to get worse; their immune systems are less effective, they are more prone to inflammation and disease, and their wounds heal slower.[50]

Hopes and despairs act directly on the healing system.[51] Feeling loved and cared for has a positive effect; feeling lonely and unloved has a negative effect. Yet few people acquire a pet simply because they hope it will improve their health by loving them. The health benefits are usually a by-product. Companion animals give us a direct, literal, living interchange with the non-human world, an interaction that many people want, perhaps without knowing exactly why.

One reason we keep other animals is because they tell us a lot about ourselves: we resonate with their animal nature through our emotional bonds with them; and we are very like them in our biochemistry, hormones, senses, and emotions such as hunger, thirst and sexual desire. We also share with them the experience of social bonds and cooperative behaviour.

In some ways, humans vastly exceed the capacities of any other animal species, as in language, culture, mythology, agriculture, art, architecture, science and technology. But in other ways, animals' senses and abilities are greater than our own, as I have said. Many non-human animals also seem to exhibit greater intuitive abilities than most humans. If we can learn from them something about these abilities, which we share to a lesser degree, we will be able to distinguish more clearly between the psychic and the spiritual realms, which are often confused.

Psychic phenomena are, as it were, horizontal, connecting animals to each other and to their environment. Spiritual phenomena are, as it were, vertical, and connect with more-than-animal and more-than-human forms of consciousness.

Both psychic and spiritual phenomena run contrary to the materialist theory that minds are confined to brains. Psychic phenomena such as telepathy indicate that minds stretch far beyond brains and can influence other minds at a distance. Spiritual phenomena like mystical experiences seem to show that minds can connect with higher forms of consciousness. Both are impossible from a materialist point of view.

That is why dogmatic materialists, like Richard Dawkins, dismiss all psychic phenomena as illusory,[52] and treat spiritual experiences as no more than subjective experiences inside brains, producing an illusory sense of connection to greater forms of consciousness that do not in fact exist.[53]

Nevertheless, there is an important distinction between the psychic and the spiritual realms. Psychic phenomena may well be widespread among animal species, and common among humans, too. But even if they are, this would not prove the existence of higher forms of consciousness transcending animal and human minds. The existence of psychic and spiritual phenomena are separate questions.

Some people accept the existence of the psychic, but reject the spiritual realm. For example, some parapsychologists are liberal materialists, and believe that psychic phenomena will eventually be explained physically, through extensions of quantum theory or other physical theories.[54] And some people accept the realm of the spiritual, but reject the psychic, including some Christians.

I myself accept the reality of both, but not because I see them as a package deal, to be accepted or rejected together. I do not see these issues as a matter of ideology, but of evidence. Psychic phenomena like telepathy are biological, part of animal nature, rather than spiritual. In order to avoid confusing these two realms, it is helpful to consider the evidence for psychic abilities such as telepathy in more detail. And there is much that we can learn about them from non-human animals.

Animal telepathy

For many years, hunters, animal trainers, horse riders, naturalists and pet owners have reported kinds of perceptiveness in animals that suggest the existence of unexplained powers. I have done much research in this area, which I have published in papers in scientific journals and summarised in my books *Dogs That Know When Their Owners Are Coming Home* and *The Sense of Being Stared At*.

There are three major categories of unexplained intuitive abilities that suggest the existence of powers, senses or fields that are not yet accepted within institutional science:

1. The sense of being stared at, knowing when another person or animal is looking at you, even if it is behind you, and silent.
2. Telepathy: picking up needs or intentions at a distance from a closely bonded member of the social group.
3. Premonitions, like the ability of many species to anticipate disasters such as earthquakes and tsunamis, sometimes more than a day in advance.

Here I discuss only one example: telepathy.

Many dogs, cats, parrots and other companion animals seem to be able to pick up their owners' thoughts and intentions, even at a distance. People's intentions, calls and commands affect their animals, and the animals' needs and emotions affect people. In random household surveys in Britain and the United States, many pet owners said that their animals were sometimes telepathic with them: forty-eight per cent of dog owners and thirty-three per cent of cat owners said that their pets responded to their thoughts or silent commands. Also, many horse trainers and riders believed that their horses could pick up their intentions telepathically.

The perceptiveness of pets usually depends on a combination of influences, including body language, familiar words, tones of voice, deviations from routine, and telepathy. Telepathic influences show up most clearly where animals pick up people's intentions and feelings when they are miles away, as in the case of telephone telepathy. In the household of a professor of ethnomusicology at the University of California at Berkeley, his wife knew when he was on the other end of the line because their silver tabby cat, Whiskins, rushed to the telephone and pawed at the receiver. 'Many times he succeeds in taking it off the hook and makes appreciative meows that are clearly audible to my husband at the other end,' she said. 'If someone else telephones, Whiskins takes no notice.' The cat responded even when her husband called home from field trips in Africa or South America at unpredictable times.

The most convincing evidence for telepathy between people and animals comes from the study of dogs that know when their owners are coming home. My colleague Pam Smart and I carried out many videotaped experiments, in which the owners went at least five miles from home and returned at non-routine times that we selected at random, travelling in unfamiliar vehicles. The filmed records showed that dogs anticipated their owners' arrivals long before they arrived home, in a way that could not be explained in terms of routine or normal sensory clues.[55] This anticipatory behaviour is common. Many dog owners simply take it for granted, without reflecting on its wider implications. Many cats also anticipate their owners' arrival, but fewer do so than dogs.[56]

Some of the most dramatic examples of telepathic links occur when people have accidents or die when they are far away from their homes. On my database, there are 140 accounts of the reaction of dogs to the death of an absent person to whom they were attached. In most cases the dogs howled for no apparent reason, but some whimpered or whined, or barked in an unusual way. In cases where no sounds were mentioned, they were said to be upset, miserable, shivering, terrified or distressed.

In most cases, the animals showed clear signs of distress at unexpected times when their people were far away, and when those looking after them had no reason to expect any problem.

A woman from St Albans, in Hertfordshire, England, was on

holiday in Ireland with her husband, when he died very suddenly on Easter Eve: 'Our seven-year-old standard poodle was staying with friends in St Albans. At just after midnight, the poodle howled and rushed upstairs to my friend, who was in the bath. At just after midnight, my husband died.'

On my database there are more than sixty accounts of cats behaving unusually when their owners died when away from home. On one occasion, a tomcat belonging to a family in Switzerland was very attached to their son, who went away to work as a ship's cook. He came home irregularly, and the cat used to wait for him at the door before he arrived. One day, the cat sat at the door and meowed very sadly. 'We could not get him away from the door,' wrote the boy's father. 'Finally we let him into our son's room, where he sniffed at everything but still continued his wailing. Two days after the cat's strange behaviour, we were informed that our son had died at exactly that time on his voyage, in Thailand.'

These telepathic connections at a distance between companion animals and people are usually related to the strength of the bond between them.[57] The love and affection between them does not necessarily imply a spiritual dimension, just as our relationships with other people may not have an explicitly spiritual aspect. But in the same way as this greater spiritual reality can shine through human relationships, so it can in our relationship with animals.

Some of the most intense experiences of unity and connection occur with horses. When riding, the rider is intimately linked to the horse and its movements both through body, emotions and intentions. Many riders are convinced that they have a telepathic bond that enables the horse to pick up their intentions. It would be difficult to investigate this experimentally, because separating telepathic effects from body movements would be difficult, if not impossible. But there is no doubt that some riders sometimes have mystical or spiritual experiences when riding.

The American veterinarian Linda Bender described a particularly memorable experience when galloping on her favourite horse:

> The joy that surged through her filled my own body as well . . .
> I was aware of our rapid breathing and of the pounding contact
> with the earth. I was totally in my body, yet able to witness it
> at the same time, because my senses had expanded to become
> one with the wind, the sun, the sky, and the Earth. There was
> no boundary, no place where I left off and the rest of the world
> began.[58]

Bender has meditated regularly for most of her adult life, but she says that her most important spiritual openings have not happened during the meditation itself: 'For me, these moments of awakening occur most often in relation to animals.'[59]

Human telepathy

Telepathic experiences reveal that we are interconnected, usually with those to whom we are strongly bonded. There are many examples of telepathy in the human realm. Perhaps the most basic and biological occurs between mothers and babies. Many nursing mothers have felt their milk let down – their breasts start releasing milk – when they are away from their baby. As a result, they go home or call home, and it usually turns out that they picked up their baby's distress exactly when the baby needed them.[60]

The most common experiences of telepathy today happen with telephones. Most people have had the experience of thinking of someone for no apparent reason, and then that person calls.[61] They say something like, 'That's funny! I was just thinking about you!' Or else people know who is calling when the phone rings before they look at the caller ID or answer it.

But is apparent telephone telepathy really telepathic? Could there be a more mundane explanation? People may think of others from time to time for no particular reason, and if someone they are thinking of then calls, this may be a chance coincidence. People may simply forget all the times they think of someone who does not ring.

The only way to resolve the question scientifically is by experiment. I have developed a simple experimental procedure to find out whether people are merely guessing, or whether their successes are more frequent than would be expected by chance. For these tests, the subjects nominate four callers they know well, usually close friends or family members. They then sit at home by a landline telephone with no caller ID display, being filmed. Then the experimenters pick one of the four callers at random, on the basis of the throw of a die or a random number generator, and call that person, asking him or her to call the subject. Thus, the subjects do not know who will be calling in any given test, because the caller is picked at random. When the phone rings, subjects have to guess who the caller is before they pick up the receiver. By chance they would be right about one time in four, or twenty-five per cent of the time.

My colleagues and I have conducted hundreds of these telephone telepathy trials. The average success rate was forty-five per cent, very significantly above the chance level of twenty-five per cent, with statistical odds against chance of trillions to one.[62] We also carried out a series of trials in which two of the four callers were familiar, and the other two were strangers, whose names the participants knew, but whom they had not met. With familiar callers, the success rate was more than fifty per cent. With strangers it was near the chance level, in agreement with the observation that telepathy typically takes place between people who share emotional or social bonds.

In addition, we have found that these effects do not fall off with distance. In some of our tests in Britain with young people who had recently arrived from the antipodes, two of the callers were new acquaintances in Britain, and the other two were family members or close friends in Australia or New Zealand. The subjects identified their distant callers better than the callers nearby, showing that emotional closeness was more important than physical proximity.[63]

Telepathy continues to evolve. Some of its recent manifestations are telepathic emails and telepathic text messages: people think of someone who shortly afterwards sends them an email or a text. My colleagues and I have done hundreds of trials on email and text message telepathy, following a similar design to the telephone tests, and with similar hit rates, very significantly above the chance level.[64]

Our thoughts and intentions appear not to be confined to the insides of our heads. They reach out far beyond our bodies. My own hypothesis is that they do so through fields that I call morphic fields.

We are used to the idea that invisible fields extend beyond material objects. The magnetic field of a magnet is both within the magnet itself and extends beyond it. The gravitational field of the earth is both within the earth and stretches out invisibly into space, holding the moon in its orbit, and holding us down to earth. The electromagnetic field of your mobile telephone is both within the phone and all around it, which is why it can transmit and receive invisible radio signals. Similarly, the fields of our minds are normally centred in our brains, but stretch out beyond them, which is why companion animals and people we know well can pick up our intentions even when we are many miles away.[65]

As I mentioned, there is an important distinction between the psychic realm, which includes telepathic communication, and the spiritual realm. Psychic connections provide important means of communication between members of animal groups, like members of wolf packs, elephant herds, flocks of birds and schools of fish.

As I show in my book, *Dogs That Know When Their Owners Are Coming Home*, there is much evidence that these psychic connections are normal, not paranormal; natural, not supernatural. They are part of animal nature, and many animals have better-developed psychic abilities than most humans. They are not in themselves spiritual, if we take the spiritual to involve a link to a higher mind or consciousness.

But these spatial metaphors are only metaphors. If the realm of the spirit includes all nature, and is present within and beyond the entire universe, it is indeed above us, extending into the sky above our heads. But it is also within us, and all around us. It is also within the animals we interact with. Linking with animals is a spiritual practice to the extent that it helps to make us more aware that there is a conscious dimension to the world that we share with all these other species.

The spiritual life of non-human animals

It is easy to assume that mystical experiences that arise through meditation or other spiritual practices are unique to human beings. Most religions take for granted that humans are superior to other animals, and this human-centred way of thinking is not confined to religions: it is even more strongly developed in secular humanism. As its name implies, humanism places the highest value on humanity. It is a form of species-ism. Other animals are of less value. Human progress comes first.

Meanwhile, in academic biology, animals are thought of mechanistically as genetically programmed machines. In this context, the idea that animals might have spiritual or mystical experiences is inconceivable.

When we think about this question in the light of research on meditation (*SSP*, chapter 1), it becomes of much more interest. What prevents humans from living in the present is the chatter of the mind, associated with the default-mode network in human brains. This is a uniquely human problem. Only humans have language and the capacity to ruminate. We may be unique among all animals in our capacity for unhappiness caused by fears and anxieties.

One reason that the keeping of pets is so popular is that they demonstrate such uncomplicated ways of being happy. A dog wagging its tail with excitement when its owner comes home is not pretending. A cat sitting on its owner's lap and purring content-edly while it is being stroked is happy. When young animals are playing they seem to be happy too, in the flow, and so do birds singing beautiful songs, or birds soaring in the sky. A lizard basking in the sunlight may be blissful.

A materialist will say that to imagine that they are happy is an invalid anthropomorphic projection of human feelings onto non-human nature. These animals are not really happy, but simply responding to pre-programmed nerve mechanisms. But materialists are also projecting. They assume that animals are machines, and project a human obsession with machinery onto them. This is arguably even more anthropomorphic and culturally conditioned, because only modern humans have machines, and all machinery is man-made.

Seeing all nature through the lens of the machine metaphor is not a vision of the truth, but a historically conditioned worldview that started in Europe in the seventeenth century and is amplified by the preoccupation with machinery in modern industrial societies.

Some diehard materialists even think of themselves as machines with no free will, and believe that their own thoughts are shaped by non-conscious brain mechanisms. To be consistent, they would have to accept that their belief in materialism is not a result of free choice, based on science and reason, but because the physical activity of their brains makes them believe that their thoughts are nothing but the physical activity of their brains.

But consistent materialists are few and far between. Most people do not think of themselves and their loved ones as machines, but as truly living organisms, as humans. And because we humans are a species of animal, *Homo sapiens*, our own feelings and experiences probably give us a better insight into the lives of other animals than projecting machine metaphors onto them. They are much more like us than like machines such as cars, jets, phones or computers.

No traditional society has thought of animals as inanimate machines, nor have Europeans traditionally done so. The very word 'animal' comes from the Latin word for soul, *anima*. In Europe until the mechanistic revolution in the seventeenth century, and practically everywhere else to this day, animals are seen as truly alive, part of a living world.[66] This attitude is often called animism, and it is undergoing a remarkable revival today under the name of panpsychism, which increasing numbers of mainstream Western philosophers and scientists are adopting as an alternative to materialism.[67]

Most societies not only see animals as truly alive, but also as participating in a spiritual reality that includes all nature. For Hindus, even today, many animal species are sacred, including cows, elephants, and monkeys. Many Hindus believe that when the god Brahma created the animals, he hid a specific secret in each of them to signify their spiritual importance to humans. Some also say that the god Shiva imparted to each of them specific states of yogic awareness.[68] Even rats are sacred; a rat is the vehicle of Ganesh, the elephant-headed god. In Rajasthan, the rat temple of Karni Mata contains some 20,000 rats, which are fed by devotees.

I discovered this surprising aspect of Hinduism when I was working in Hyderabad. I lived in the wing of a crumbling palace, and one night I woke up in horror to find a rat walking across my face. The next morning, I asked my Hindu cook, Swamy, to buy a rat trap and try to get rid of the rats. The following day when I came home from work he showed me a live rat inside the cage-like trap, and I asked him to kill it. He said he couldn't, because it was a sacred animal and killing it would bring bad luck. Instead, he put the cage containing the rat on the back of his bicycle, took it several miles away, and released it.

I subsequently noticed men on bicycles all over Hyderabad with rats in cage traps taking them away to release them. The total number of rats remained the same, but they were continually relocated.

In Europe in the thirteenth century, the theologian St Thomas Aquinas summarised a Christian way of looking at animals and other creatures as follows:

> [E]very movement and activity of any creature seems to tend towards something perfect. But the perfect has the nature of good, since the perfection of anything is its goodness. Therefore, every movement and action of any creature tends towards something good. But any good is a likeness of the highest good, just as any existing is a likeness of the first being. Therefore, the movement of action of anything tends towards assimilation of the divine goodness.[69]

Animals are aspiring to divine perfection. How different from the standard modern view! According to Aquinas, animals are participants in the divine being, and their happiness participates in divine happiness.

What we can learn from animals

Animals can help us in many ways, some practical, some emotional, and some spiritual. Here I summarise some of the spiritual lessons we can learn from them. By spiritual, I mean a flow of consciousness that connects us to more inclusive, higher forms of consciousness, and even to the source of consciousness itself. Some of these spiritual lessons give us a more realistic sense of ourselves than the narcissistic assumption that we are the masters and possessors of nature, which is ours by right, and consists of machinery that we can control.

1. Non-human animals can teach us humility. Many animals have sensory abilities that go far beyond our own, including the ability of dogs to hear noises far above our range of hearing, and badgers to smell with a much greater sensitivity than our own. Eagles can see small animals like mice from a height of hundreds of feet, four or five times further away than a human could spot them. Bees' vision works in the ultraviolet, far beyond our visual range. Sharks detect electric fields of which we are unaware. Bats fly fast and catch their prey on the wing by echolocation.

   Similarly, many animals have physical skills that far exceed our own, like the ability of gibbons to leap from tree to tree, of buzzards to soar on thermal currents; moles to burrow through the ground, finding their way in total darkness; and sea turtles to migrate over thousands of miles to the island where they were born. Some kinds of animals, like tardigrades, are much tougher than we are in the face of extreme heat, cold, pressure, radiation and dehydration that would quickly kill a human. We have intellectual, technological, linguistic and other abilities that other animals do not; but other species have abilities that far exceed our own, even when we are aided by technology.

2. Unconditional love. Many religions encourage unconditional love or compassion as a virtue, and see the source of love in the ultimate spiritual reality, whether called God or by another name. Many humans show unconditional love for other humans,

as many parents do for their children, and saints show some of the most outstanding examples of selfless love. Members of many animal species also show unconditional love, and are prepared to die to save their young or other members of their group. Among the social insects – ants, wasps, bees, and termites – most are prepared to lay down their lives for the sake of the society as a whole. And when animals form bonds with us, they often show an unconditional love towards us.

We can choose to see this behaviour as purely instinctive, programmed by selfish genes for selfish ends. But the love that can flow through animals towards us, and from us towards them, may not just involve the animals' and our emotions, but may be part of a greater flow of love to which we and animals are both connected.

3. Healing. Many companion animals have a healing presence. This effect depends on their good intentions and their love. Humans can have a healing presence, too. Animals show us that being present in a loving way with someone in need can help that person feel better emotionally, and can also help the process of healing. We can call this an aspect of the placebo effect if we like, but placebo effects work.

4. Many animals show a greater intuitive sensitivity than humans. They are sensitive to each other, and when they live with humans some are sensitive to human emotions, intentions and needs, even at a distance. Their telepathic abilities remind us that our own intuitions are part of our animal nature, although we often ignore them. Animal premonitions – like animals seeming to anticipate earthquakes and tsunamis – also remind us that many animals have ways of knowing that we have forgotten, or that have fallen into disuse.

5. We are often distracted by thoughts, worries, ruminations, resentments and fantasies. In order to bring us back into the present, we can adopt spiritual practices such as meditation. But many animals live in the present without these distractions as a normal state of being. Most non-human animals live more intensely in the present than most humans. They participate more fully in what is happening. We can learn from them how to be more present.

6. Dying. When we live through the deaths of animals we love, we learn about loss and grief – and that life goes on. For many children, the death of a beloved pet is their first encounter with loss and grief.

7. Many animals seem to experience joy or bliss through play, through flying or soaring, through swimming, through jumping, through moving gracefully, and through song. They remind us that joy is inherent in the world, not just in us. Our joy and the joy of other species have a common source. In the Hindu conception, ultimate reality is *sat-chit-ananda*, being-consciousness-bliss. *Sat-chit-ananda* underlies humans and all other animals, and all nature, and one of its aspects is joy.

Two practices with animals

BE PRESENT WITH AN ANIMAL

If a cat is purring while you stroke it, be completely present to the stroking and the purring – rather than stroking distractedly while having a conversation or watching TV. The cat is present; become present with it.

Or listen to a bird singing. I live in England, and my favourite birdsong is that of blackbirds singing in the spring and early summer. I listen to their songs, which change every time they sing.

Often I hear another blackbird respond: they interact with each other, and reply to each other's tunes and variations. They are present to each other. We can be present through listening. Wherever you live, you will be able to find birds singing.

And there are many other ways to be present with animals. Unlike us, they are not distracted by internal dialogues. We can learn much from their quality of presence.

GET TO KNOW ANOTHER SPECIES

If you keep a cat, dog, horse, parrot, budgerigar, rabbit, hamster, ferret, lizard, goldfish, stick insect or another kind of animal, you are already getting to know another species. If you have, or have had, more than one cat, dog, horse or other animal, you will also know that each animal is different. Each expresses its unique individuality within the context of its species' instincts.

If you do not have a companion animal, or even if you do, you can get to know a wild species by observing individuals that live near you – like birds in your garden or in a nearby park – watching and listening to them, perhaps feeding them, relating to them throughout the year. Or you can raise caterpillars or tadpoles and witness their transformation into butterflies, moths or frogs.

The better you know your chosen kind of animal, the more you will appreciate its way of being, its form of life. You will feel connected to a world much wider than your human concerns, and with which you share a common source.

3
Fasting

Fasting is a voluntary abstinence from food or drink or both. It plays a part in practically all religious traditions, and it has now been secularised. You can pay large sums of money to eat almost nothing at health farms and clinics. Rebranded as detoxification, the purpose of fasting can be seen as purely physical, as opposed to spiritual. But fasting is essentially free and indeed saves money.

In this chapter, I discuss both total and partial fasting, and their evolutionary, cultural and religious backgrounds. Fasting even has a political dimension, as in hunger strikes. But its traditional purpose is as a spiritual practice.

The anthropology of fasting

In many cultures, fasting occurs in a variety of ritual, seasonal and initiatory contexts. People fast as an act of penitence or purification; as a preparation before a festival or rite of initiation; as part of a mourning ceremony; as a means of inducing dreams and visions; or as a protest, as in hunger strikes.[1]

In many cultures, mothers are expected to abstain from certain foods before or after the birth of a child. For example, in New Britain, an island in Papua New Guinea, pregnant women could not eat cuttlefish, which was believed to walk backwards, in order that their children should not become cowards.[2] Some of these taboos may have been entirely magical in their basis, but others may have had good health reasons, as in contemporary recommendations to pregnant women not to drink alcohol or take recreational drugs to avoid harming their baby.

Fasts were also an important part of rites of passage. Among some First Nation tribes of British Columbia, girls were put into seclusion after their menstrual period, and fasted for four days. Likewise, many Native American boys spent a period away from human habitations while they fasted as part of a vision quest. Among the Algonquin, 'they begin by blackening the boy's face, then they cause him to fast for eight days without giving him anything to eat.' During this period, his dreams were carefully enquired into. In one of these dreams, the boy might learn what his 'medicine' was to be. Likewise, in classical Greece and Rome, fasting often preceded initiations into mystery cults.

In the Old Testament, when Moses received the Ten Commandments on Mount Sinai, he fasted forty days and forty nights (Exodus 34: 28). The Prophet Elijah fasted for forty days and forty nights on Mount Horeb (1 Kings 19: 8). Likewise, Jesus fasted for forty days and forty nights in the wilderness, after his baptism in the River Jordan by John the Baptist (Matthew 4: 1–2).

In many cultures, people fast in order to induce dreams or revelations from the spirit world. Among Native Americans, fasting was an important part of the preparation for becoming a medicine man or shaman, and was a common practice for acquiring hidden knowledge or messages from spirits in dreams. Hunters fasted until they dreamed whether their hunts would be successful or not. Husbands fasted until they dreamed whether or not their hopes of becoming a parent would be gratified. 'The greater the power of fasting, and the more vivid and numerous the consequent dreams, the more was the seer held in reverence and the greater power did he acquire.'[3]

The value of fasting for visionary experience was widely accepted in the early Church. St John Chrysostom (c. 349–407) said that fasting 'makes the soul brighter and provides it with wings to mount and soar.'[4]

In cultures where fasting is part of the process of mourning, the fasts are usually brief. Among the Yoruba in West Africa, widows and daughters were expected to refuse all food for at least one day. And in many cultures a period of fasting preceded a funeral feast.

More generally, fasting was often part of a rite of preparation for a feast or a ceremony. In ancient Egypt, people fasted and performed ablutions before entering a temple. In North America, among the Natchez, at the Festival of New Fire, a harvest festival, the people fasted for three days before the festival began. In ancient Greece, the celebration of the Mysteries in the cave of Eleusis was preceded by a fast of nine days before eating and drinking the sacramental food.

A contemporary version of this preparatory fasting is still found in the Roman Catholic Church, where many people traditionally fast before receiving Holy Communion. And in the traditional Catholic and Anglican calendars, all the major feast days are preceded by fasts, usually short fasts on the eve of the feast. The most important feasts, Christmas and Easter, are preceded by the longest periods of abstinence or fasting, namely the seasons of Advent and Lent.

The Lenten fast, which begins on Ash Wednesday, is not only followed by a feast, Easter, but also preceded by a feast, on Shrove Tuesday, or Mardi Gras, or Carnival, celebrated vigorously in Brazil and elsewhere. The traditional meaning of carnival is about giving up meat, from the Latin *carne* = meat and *vale* = farewell.

Penitential fasting is generally combined with prayer, and in some cultures is believed to make gods or spirits more likely to respond. In the ancient world, in Egypt, Babylon and Assyria, people often fasted in periods of distress or calamity, praying for forgiveness and mercy. And individuals often fasted as an act of penance for their own sins. In the Jewish tradition, this form of penitential fasting was, and still is, ritualised in the fast of the Day of Atonement, Yom Kippur.

In Islam, the Koran recommends penitential fasting for three days on a pilgrimage and for seven days on returning (Surah 2: 196). A believer who kills another and cannot find the blood money to recompense the family of the victim must fast for two months as a penance (Surah 4: 92). The oath breaker who cannot feed ten poor men as a penance must fast for three days (Surah 5: 89). And all Muslims are expected to fast during the month of Ramadan, abstaining not only from eating, but also from drinking anything between dawn and dusk (except for the sick and infirm, travellers and children). The inner meaning of this practice is detachment. In one of the *hadith*, or sayings of the Prophet Mohammed, God says, 'All the deeds of a human being are for himself [or herself], but fasting is solely for Me, and I shall reward him [or her] for it.'[5]

Buddhist monks traditionally eat only one meal a day, at midday, and thus spend almost twenty-four hours in fasting on a regular basis. They also fast on the days of a new and the full moon, together with a confession of their sins. The anniversary of the Buddha's death is traditionally preceded by a five-day period of abstinence, in which lay people also share.

In Hinduism, there are many forms of religious fast. Many people fast at least one day a week, and there are also monthly and yearly fasts like the great Shiva Festival of Maha Shivaratri, with fasting through an all-night vigil and on the subsequent day.

Fasts also take on a political significance as a form of persuasion or protest. In India, the first recorded example of a threatened hunger strike is in the *Ramayana*, an epic that dates from at least 500 BC. The god Bharata threatens to fast in front of his brother Rama, in order to persuade him to return from exile to rule his kingdom. The use of fasts as a public call for justice, usually carried out at the door of an offending party, was such a troublesome practice in India that the Government passed a law against it in 1861.[6] But hunger strikes continued to be an important form of non-violent protest, and were used to great effect by Mahatma Gandhi when the British jailed him.

In pre-Christian Ireland, as in India, there was a traditional legal remedy of 'fasting against' a person. The person who fasted made a request, and then proceeded to fast. If the fasting led to the faster's death through the persistent refusal of his request, then the death of the faster brought his blood upon the head of the person he had fasted against.[7]

As in India, this ancient practice was used to great political effect in the twentieth century, most notably in 1981, when ten republican paramilitaries imprisoned by the British in Northern Ireland highlighted their cause by a hunger strike. The men lived without food between forty-six and seventy-three days.[8] The first to die was Bobby Sands. Their deaths provoked serious public disorder, and the government was forced to make concessions.

Although there have been many traditional religious, cultural and political motives for fasting, it was not until the nineteenth and twentieth centuries that it was portrayed as a purely secular method for improving health. Fasting cures were promoted not only by enthusiastic doctors and therapists, but by popular writers such as Upton Sinclair. These were for overweight people, but also relieved a wide variety of ailments. Sinclair himself undertook intermittent fasts of up to twelve days.

As he wrote in his book, *The Fasting Cure* (1911):

The reader may think that my enthusiasm over the fasting cure is due to my imaginative temperament; I can only say that I have never yet met a person who has given the fast a fair trial and who does not describe his experience in the same way . . . I

regard the fast as Nature's own remedy for all other diseases . . .
And I believe that when the glad tidings of its miracles have
reached the people, it will lead to the throwing of ninety per
cent of our present *materia medica* into the wastebasket.[9]

Naturopaths and several other kinds of therapists have long advocated fasting, and some run clinics where people can fast under
supervised conditions. But in general, the mainstream medical
profession has ignored its many benefits, which I discuss below.
Most doctors have no personal experience of fasting and treat it
with suspicion. In the modern secular world, fasting is probably
practised less than ever before. Meanwhile the incidence of obesity
and type 2 diabetes has soared.

Evolutionary responses to starvation

Fasting differs from starvation, because it is voluntary rather than
involuntary, but the ability to survive a temporary lack of food is
deeply embedded in human and animal nature. Food supplies are
intermittent under natural conditions. Even when there is a large
amount of food, a period of shortage often follows.

These are the conditions under which many kinds of organism
have evolved, including microbes. And intermittent food supplies
are often beneficial. When the bacterium *E. coli* is transferred from
a nutrient-rich broth to a calorie-free medium, the starved cells
survive four times longer than sister cells that remain in the nutrient
broth. In yeast, when cells are deprived of food by transferring
them from a solution of nutrients to water, their lifespan doubles,
and they have an increased resistance to stress.[10]

The same principles apply to many multicellular animals. When
the nematode worm *Caenorhabditis elegans* is given very little
food, or intermittently starved, its life span almost doubles. In
fruit flies, reducing the food supply also extends life span. And so

it does in mice subjected to an alternate-day fasting regime. In one experiment, mice that were fed only on alternate days lived almost a year longer than similar mice that were fed every day, a thirty per cent extension of their lifespan.[11]

Intermittent fasting also led to a major reduction in the incidence of cancers. Even when they were living in the presence of carcinogenic chemicals, fasting seemed to have cancer-preventing effects in mice.[12] Humans, too, have evolved under conditions of irregular food supply, and an assured sufficiency of food every day is a recent luxury. Even in the twenty-first century, there are still many millions of people with inadequate and irregular food supplies.

Physiological effects of fasting in humans

The physiology of fasting revolves around glucose, insulin and fat. Our brains need glucose, as do other cells in our bodies. Glucose is carried throughout the body in our blood. When we eat carbohydrate foods, the level of glucose in our blood goes up, rapidly if we eat sugar and slower if we eat starch, which takes longer to digest. The rise in glucose levels triggers the secretion of the hormone insulin by the pancreas, and as the insulin circulates in the blood, it stimulates cells to take up glucose. In the muscle cells, this sugar can be used to provide muscular energy. In the liver, glucose molecules are strung together into glycogen. Glucose taken up by fat cells is turned into fat, which is the body's principal way of storing energy.

The store of glycogen in the liver lasts only about 12–16 hours when we are not eating, but we are capable of accumulating enough stored energy in fat cells to last for weeks, even months. There are fat cells around the internal organs, the buttocks, and all the other parts of the body that enlarge as people become obese.

When we are fasting, we rely primarily on stored fats. The proteins in the muscles are broken down very slowly by comparison. As we break down fats, much of the fat is turned into chemicals called ketones, like acetoacetic acid, beta-hydroxybutyric acid and their breakdown product, acetone (Figure 1). These circulate in the blood, and the ketones are absorbed by cells and can be used for the production of energy. Some are converted into glucose in the liver, helping to maintain the glucose levels in the blood.

KETONE BODIES AND RELATED COMPOUNDS

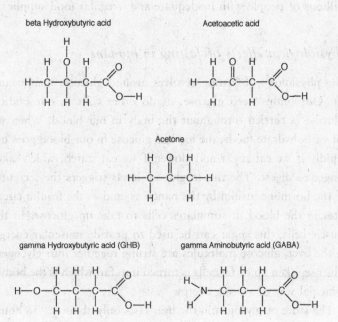

Figure 1. **Ketone bodies and related compounds.**
During fasting and on a ketogenic diet, human blood and urine contain three main 'ketone bodies' as shown in the upper part of this figure: beta Hydroxybutyric acid, acetoacetic acid and acetone. Below, the structures of gamma Hydroxybutyric acid (GHB) and gamma Aminobutyric acid (GABA), both of which are neurotransmitters.

These ketones have a characteristic sweetish smell, which can be detected in the breath and urine of people who are fasting. This condition is called ketosis.[13] Fat stored in the body is used as a source of energy in people who are starving or fasting. And ketones affect neurotransmitters, brains and minds, as I discuss below.

Most people have enough fat reserves to enable them to go entirely without food for many days, and those who are very obese have enormous energy reserves. In one clinically controlled test, an obese young man in Scotland went without food for 382 days with no ill effects.[14] Indeed there were good effects, because he lost a lot of weight. Even people who are not obese have enormous reserves of energy in the form of fat, and could survive without eating for several weeks. In theory, most Americans could walk from New York to Florida without eating anything at all, simply by using the fat reserves in their bodies.[15]

During fasting, the pituitary gland secretes increased amounts of growth hormone. As its name implies, this hormone plays an important role in the normal growth and development of children and adolescents. The levels of growth hormone thereafter decrease with age. Meals suppress the secretion of growth hormone, and overeating suppresses it even further. Very low growth hormone levels in adults lead to more body fat, less muscle mass and decreased bone density. Some of the effects of ageing seem to result from low levels of growth hormone. Injecting people with growth hormone increases the mass of muscles and bones and decreases the amount of fat. But such injections have undesirable side effects, including increasing the levels of blood sugar, increasing blood pressure and leading to more prostate cancer and heart problems.[16] This is why injections of growth hormone are rarely used.

The most powerful way to stimulate the secretion of natural growth hormone without these undesirable side effects is by fasting. In one study, growth hormone secretion more than doubled over a five-day fasting period, and a study of a religious forty-day fast found that growth hormone levels increased more than tenfold without the use of any drugs at all.[17]

Most people who undertake prolonged fasts find the first day or two most difficult because they feel waves of hunger. This problem becomes less from the third day onwards. Also, those who have not fasted before may suffer from headaches during the first day or two, as toxic substances are flushed out of the fat cells. Again, things improve thereafter.

For more than twenty years, I have made a practice of fasting for between three and seven days during Lent, usually in the week before Easter, Holy Week. I do not eat anything, and drink only water, tea and coffee. The first year I had a serious headache on the first and second days, but not in the following years. Like many other people, I found that by the third day I stopped feeling hunger pangs. Not eating became easier.

However, when I fast, the days seem to be longer than usual, without breaks for meals, and food smells extraordinarily delicious. But I can concentrate better, and have more vivid dreams.

Obesity and diabetes

These basic physiological facts help explain the worldwide epidemic of obesity and diabetes. According to the World Health Organization, in 2014 more than 1.9 billion adults were overweight, thirty-nine per cent of all adults, of whom 600 million were obese. In addition, 41 million children under the age of five were obese or overweight, nearly half of whom were in Asia.[18] As well as an increased chance of premature death when they grow up, obese children often have breathing problems, have high blood pressure and show insulin resistance, making them prone to type 2 diabetes.

The number of people suffering from diabetes quadrupled between 1980 and 2014, when it affected eight and a half per cent of all adults. Consequently, many more people suffered from diabetes-induced blindness, heart attacks, kidney failure and lower limb amputations.[19]

Type 1 diabetes has unknown causes, and is a result of deficient insulin production by the pancreas. It can be controlled by daily administrations of insulin. By contrast, type 2 diabetes is largely associated with excessive body weight and physical inactivity. The pancreas produces insulin in response to sugar in the blood, but the insulin stops working effectively because the body is so often flooded with it as a result of frequent meals, snacks and sugary drinks. In type 1 diabetes, there is too little insulin to cause cells to take up sugars and reduce the level in the blood. In type 2 diabetes, there is plenty of insulin, but the cells have become resistant to it.

For many years, official nutritional advice was based on the assumption that what made people fat was eating fat. They were advised to switch to low-fat diets, which usually resulted in their eating more carbohydrates.

Eating three or more meals a day with sugary drinks and frequent snacks creates the perfect conditions for weight gain and for type 2 diabetes. Exercise helps to reduce some of these problems, but switching to a low-carbohydrate diet is the most effective way of losing weight and overcoming insulin resistance. There are now many low-carb diet systems, including intermittent fasting. But if dieters treat their dieting as a temporary phase and then go back to their previous eating habits, they usually put on weight again. To lose weight and to keep it off involves a change in lifestyle, which may include missing out a meal, such as breakfast, and avoiding sugar and starchy foods.

Fasting is a simple and often effective way of reducing the effects of type 2 diabetes, or even curing it. When people are not eating, insulin secretion is no longer stimulated. Insulin levels in the blood drop, and cells regain their sensitivity to insulin.

Effects of ketosis on brains

Brains normally rely on glucose as their energy source, but in conditions of ketosis, they use ketones instead. Ketosis occurs not only in fasting and starvation, but also with diets high in fats and low in carbohydrates.

In ketosis, one of the ketones in the blood, betahydroxybutyric acid (BHB), is very similar chemically to gamma aminobutyric acid (GABA) (Figure 1), which is the most important neurotransmitter in the brain. About forty per cent of the synapses in human brains work with GABA. The GABA receptor molecules in nerve cell membranes are called channel receptors, because when GABA binds to them they change shape slightly and allow chloride ions to enter through this molecular channel in the membrane to the inside of the cell. As a result, the negatively charged chloride ions reduce the excitability of the cell membrane.

Thus GABA has an inhibitory, rather than an excitatory effect. Some medications have molecules that bind to the GABA receptors, enhancing the natural effects of GABA, particularly benzodiazepines such as Librium and Valium, which are used to alleviate anxiety.

BHB leads to increased levels of GABA in brains, and has several effects of brain activity. For more than a century, high-fat diets have been used medically to produce ketosis as a way of reducing the frequency of seizures in people with epilepsy.[20] GABA is sometimes called a 'feel-good' neurotransmitter, and increased levels of GABA may well underlie some of the beneficial effects of ketosis in reducing anxiety and stress, and giving better mental focus.[21]

Most clinical studies of ketosis have been carried out in the context of controlling epileptic seizures. But probably some of the subjective effects of fasting, including enhanced mental clarity, reduced anxiety and euphoria, are also a result of the effects of BHB on GABA levels and on GABA receptors.[22]

A compound closely related to BHB is gamma hydroxybutyric acid or GHB (Figure 1). GHB occurs naturally within the brain and is a neurotransmitter in its own right, as well as enhancing the effects of GABA. In the 1960s, it was used medically as an anaesthetic. It is currently used as a drug for the treatment of narcolepsy and, in some countries, in the treatment of alcohol dependence and withdrawal.[23]

It is also psychoactive, and has become a drug of abuse in Europe because of its effects in inducing euphoria, tranquillity and increased sex drive. It is often referred to as a 'club drug' or 'date rape' drug.[24] BHB may act in similar ways to GHB on neurotransmitters, helping to explain the subjective effects of fasting. But whereas GHB is addictive, the effects of ketosis induced by high-fat diets or by fasting are not.

In clinical studies, some of the effects of fasting included improved memory, a sense of being energised and revitalised, a feeling of wellbeing and euphoria.[25] Some of these may well result from the psychoactive effects of BHB.

Dangers

Fasting is not advisable for children, for pregnant women, for people suffering from anorexia or for people with gout. It can also be dangerous for some people on medication, especially anti-diabetic medication, because the changes in blood sugar and insulin level caused by fasting may cause the medicine to work inappropriately.

For normal, reasonably healthy people, fasting is rarely dangerous, even for periods of a week or more. But for people with medical problems, it should be done under the supervision of an experienced person. Unfortunately, most doctors have little experience of fasting and are not taught about it in medical schools. However, some doctors have become expert guides, as are most naturopaths.

Fasting to death

When I was living in Tamil Nadu, South India, some friends took me to visit an ancient cave temple complex, at Sittanavasal, built by Jains in the second century AD on an isolated rocky hill. For me, the most interesting and surprising feature of this complex were the Jain beds, polished human-shaped indentations in the rock, dating back to the first century BC, within a natural cave. I learned that these were the places where Jain monks fasted to death when they felt they were nearing the end of their lives. Some were commemorated by inscriptions.

For years I assumed that this was an ancient practice that had died out, until an Indian friend, Satish Kumar, told me about the death of his mother. Satish's family are Jains from Gujarat, in North India. When his mother was old and felt she was nearing the end, she went round the other families in the village to say goodbye, and called her own family to be with her. She then fasted to death and took leave of this life in peace.

I now realise that both the ancient Jain monks and Satish's mother were taking part in an ancient practice called *sallekhana* or *santhara*, highly respected within the Jain community, but permissible only in the event of unavoidable calamity, distress and age. Being gradual, it gives the dying person ample time to reflect on his or her life, to achieve appropriate closure with family members and with the wider community, and to die peacefully.

A similar practice of fasting to death in Hinduism is called *prayopavesa*; only a person whose death is imminent with no remaining responsibilities is entitled to perform it. In a more extreme practice, in Japan some Buddhist monks of the Shingon school underwent a very slow practice of starvation and meditation called *sokushinbutsu*. This process could take up to ten years before they died in a state of internal mummification.[26]

Jain texts and traditions make it clear that *santhara* is not suicide, but rather a way of choosing to live with dignity until death. In 2015, secular activists challenged this tradition in the courts, arguing that it contravened the laws against suicide in India. The High Court of Rajasthan upheld their argument and banned the practice, but following nationwide protests by Jains, the Supreme Court of India lifted the ban.[27]

These age-old practices and modern legal arguments in India take on a much wider significance in twenty-first-century Europe and North America, where there is an intense debate about assisted dying. In some countries, doctors, nurses or family members who assist terminally ill people to die by administering drugs can be prosecuted for murder. Yet many people who are suffering from painful or severely distressing medical conditions would like their death to be accelerated.

Unlike administering drugs to hasten death, terminal fasting is legal and is already widely practised, though rarely publicised. However, in 2014 this issue was widely discussed in the United States when Diane Rehm, a well-known radio host, revealed that her husband had hastened his death by stopping eating and drinking when he was terminally ill with Parkinson's disease. In the medical literature this is known as VSED, voluntarily stopping eating and drinking.[28]

Many people lose their appetite as they near the end of their lives, and by choosing to fast they may be helping a natural process. In one small study in Oregon, where doctor-assisted suicide is legal, more people chose death by fasting than by a prescription drug. US courts have affirmed the right of patients to refuse medical treatments, and many doctors and medical ethicists believe that this includes the right to fast and to refuse force-feeding, as long as people are capable of making the decision for themselves.[29]

Living without food for years

Not all prolonged fasts result in death. I first heard of this phenomenon when my wife and I visited Jodhpur, in Rajasthan, India, in 1984. An Indian friend took us to visit a local holy woman called Satimata in the nearby village of Bala. We were told that when her husband died in 1943, when she was about forty years old, she wanted to immolate herself on his funeral pyre in the tradition of *sati*, but she was prevented from doing so. Instead, she vowed never to eat again. Having been prevented from dying by fire, she intended to die by fasting. She did not.[30]

When we met her, she was supposed to have lived for forty-one years without food or drink, and without producing faeces or urine. Yet she looked like a normal, elderly village woman, apart from the fact that she was surrounded by devotees. When we were there, she had a cold and had to blow her nose several times; she seemed to be defying not only the law of conservation of energy, but also the law of conservation of matter, generating mucus but taking in no food or water.

I assumed that she must have been eating and drinking secretly, but her devotees were adamant that she was genuine. Some had known her for years, even lived with her, so had the opportunity to see if she was eating behind the scenes. And when I met her, and talked to people who knew her, she did not strike me as a charlatan, but as a woman of sincere faith.

I later found that Satimata was not unique; there were other holy women and men in India who were supposed to have lived without food for years. Some had been exposed as frauds, but others had been investigated by medical teams who found no evidence of secret eating.[31]

The scientific term for this phenomenon is *inedia*, from the Latin word for fasting. In India, the explanation most commonly advanced for this apparent ability to live without food is that the energy is derived from sunlight or from the breath, and in particular from *prana*, a life force in the breath. This is why some people who claim to live with little or no food call themselves breatharians.

In the West, there have also been many claims that people can live for long periods without eating, including holy men and women like St Catherine of Siena (d. 1380); St Lidwina (d. 1433), who was said to have eaten nothing for twenty-eight years; the Blessed Nicholas von Flüe (d. 1487), nineteen years; and the Venerable Domenica dal Paradiso (d. 1553), twenty years.

In the nineteenth century, two saintly women were said to have eaten nothing for twelve years, except consecrated wafers in Holy Communion: Domenica Lazzari (d. 1848) and Louise Lateau (d. 1883).[32] In the nineteenth century there was also a widespread 'fasting girl' phenomenon in Europe and the United States. Some may have been anorexic; some were exposed as frauds; but there are also some well-documented cases where girls lived for years without eating.

Herbert Thurston, a Jesuit scholar, documented this fascinating phenomenon in his classic study, *The Physical Phenomena of Mysticism* (1952). He pointed out that not all cases of inedia occurred in particularly spiritual people. For instance, a Scottish girl, Janet McLeod, who seemed to survive without food for four years, was thoroughly investigated by doctors, and the case was reported in the *Philosophical Transactions of the Royal Society* in 1767. This young woman was seriously sick rather than saintly.

Also in the eighteenth century, Pope Benedict XIV asked the medical faculty of the University of Bologna to investigate cases of inedia. In their report, while fully recognising the likelihood of imposture, credulity and mal-observation, the doctors upheld 'the genuineness of certain well-attested examples of long abstinence from food though no supernatural causation could reasonably be supposed.'[33] As in the case of Janet McLeod, some of these cases seemed to result from illness.

The best-documented example in the twentieth century was the Bavarian mystic Therese Neumann (1898–1962), who stopped eating solid food in 1922. She attracted much public attention, and the Bishop of Regensburg appointed a commission to investigate the case, headed by a distinguished doctor. Therese was closely observed for two weeks by a team of nursing sisters, who were confident that during that period she did not take either food or drink.[34] But, as Thurston acknowledged, no amount of evidence would alter the opinions of committed sceptics, who declared her to be 'a vulgar imposter'.

After considering many religious and non-religious cases, Thurston concluded, 'We are forced to admit that quite a number of people in whose case no miraculous intervention can be supposed, have lived for years upon a pittance of nourishing food which can be measured only by ounces; and upon this evidence we shall be forced to admit the justness of the conclusion of Pope Benedict XIV that mere continuation of life, while food and drink are withheld, cannot be safely assumed to be due to supernatural causes.'[35]

If a Pope and a leading Jesuit scholar favour a natural rather than supernatural explanation, what might it be? No one knows. Clearly, inedia is rare, and in most cases prolonged fasting leads to death, as in the case of people who go on hunger strike or practise *santhara*. But the persistent reports of inedia show that more is going on than either common sense or conventional science would lead us to expect.

Secular and spiritual fasts

Fasting can be carried out for non-religious reasons, to lose weight, to detoxify the body, to help prevent diabetes, or to achieve better health. Or it can be carried out for religious or spiritual reasons, as a practice of self-discipline, and as a way of intensifying the ability to pray and to seek spiritual guidance.

In all traditions, fasting has been experienced as an aid to prayer; in some traditions it is believed to make prayers more powerful. Similarly, fasting to death can be done in a secular spirit as a way of hastening the end of a terminal illness, or as a spiritual practice, as in the Jain *santhara*. Even living without food for long periods, inedia, can occur in saintly people and in people who are not saintly but sick. But often these secular and spiritual aspects of fasting go together. Even if a person's motive is to promote physical health, the fast may bring about a new kind of mental clarity and self-understanding, with unexpected spiritual benefits. Conversely, people who fast for spiritual reasons are likely to enjoy better physical health.

Two ways of fasting

FOODLESS FASTING

Just try it. Give up eating, and drink only water or other non-calorific drinks like tea or herbal tea. If you have not fasted before, you may experience a headache by the end of the first day, and you will probably feel waves of hunger over the first two days. But if you continue for three days or more, it gets easier. A three- or four-day fast is a realistic objective, but if you want to feel your way into this more gradually, you could start with one- or two-day fasts.

Because no food is going into the gut, much less comes out, and you will probably not need to go to the lavatory much during your fast. But to avoid constipation, I find it helps to take a daily source of fibre while I am fasting, in the form of capsules of psillium husks. I also take a daily multivitamin and mineral pill to avoid vitamin or mineral deficiencies.

You can, of course, fast at any time, but you may find it more effective if you do it during a traditional fasting period like Lent, if you come from a Christian background. The fact that you are placing yourself in a long tradition, and that millions of other people are observing this period of fasting all around the world, makes it easier – at least, it does so for me. And by doing it within a spiritual context, you may find that it makes other spiritual practices like prayer and meditation more effective – which is, after all, one reason why fasting is part of all religious traditions.

Before embarking on a fast, make sure that you are unlikely to be at risk (see the list of dangers on page 85), and if you are in doubt, seek the help of an experienced advisor, such as a naturopath. It would also be a good idea to read a book on the subject, such as Jason Fung's *Complete Guide to Fasting*.[36]

A PERIOD OF ABSTINENCE

In many religious traditions, fasting seasons are not solely for total fasts, but for a period of abstinence from particular kinds of food or drink, as when Christians traditionally give up meat for Lent, eating fish or vegetarian food instead.

When I was living in India, I followed this traditional practice and became a vegetarian during Lent. I found it suited me well, and stayed more or less vegetarian thereafter. So the following year I gave up alcohol instead, which I have now done every year for more than forty years. Some people give up smoking tobacco or cannabis, or both; some give up sweets or chocolate; some give up fast foods; and some give up patterns of behaviour like complaining. There is plenty of scope for choice.

The stronger the habit that this period of abstinence breaks, the harder it is to do, but the greater the benefit. However, it is also made easier by the fact that the Lenten fast is forty days long, whereas the period from Ash Wednesday to Easter Sunday is forty-six days. The reason for this difference is that for Christians, all Sundays are feast days, and so the six Sundays that occur in this period are excluded from the fast.

If you prefer a secular approach, then you could choose a

calendar month as a period of abstinence. Perhaps the most common month to do this is January, following a New Year's resolution.

If you come from a non-Christian background, then observe a fasting period in your own tradition. And as well as longer periods of fasting, in all traditions there are also fasting days, like Fridays for Christians, that provide weekly opportunities for a day of abstinence or fasting.

4
Cannabis, Psychedelics and Spiritual Openings

Consciousness-altering drugs have been used in shamanic and religious traditions for thousands of years. Traditional mind-altering herbs, brews and mushrooms are now widely available, even if illegal, in many modern cities, together with a wide range of synthetic psychoactive drugs.

What is the nature of the experiences these substances induce? Can they connect human minds with other forms of consciousness? Can they open the way to genuine spiritual experiences? These are the principal questions I explore in this chapter.

The most common mind-altering drug is alcohol, which occurs naturally in yeast-fermented fruits. Monkeys, like many other species, have a low tolerance for alcohol.[1] Humans have a higher tolerance, largely as a result of an enzyme called alcohol dehydrogenase-4 found in the liver. This enzyme seems to have arisen about 10 million years ago in apes that were ancestors of humans, gorillas and chimpanzees, enabling them to eat fermented fruit without being incapacitated by drunkenness.[2]

The cultural use of alcohol goes back thousands of years. Fermented drinks based on honey probably came first; many ancient texts suggest that mead long predated beer and wine.[3] Ancient beers were created independently around the world as long as 30,000 years ago. Many contained medicinal herbs.[4] Wine came later. Today various forms of fermented drinks are found in many cultures, and alcohol has a religious sanction in many traditional societies.

In ancient Egypt, beer and wine were offered to the gods in

rituals, as they were in Sumeria and Babylon. In ancient Greece, wine played a central role in the Dionysian festivals and rituals. For Jews, wine is an integral part of the ritual celebration of Passover. For Christians, it is an essential part of Holy Communion, a ritual re-enactment of Jesus' Last Supper with his disciples, which was itself a Passover dinner.

Cannabis has a long history of use in Asia, including by Indian holy men or *sadhus*. In ancient Greece, the Eleusinian mysteries included an initiatory rite involving a mind-altering brew. The ancient Vedic texts of Hinduism include hymns to *soma*, a transformative drink made of fermented plant juice, whose identity is unknown.

In Central and South America, a number of traditional cultures ritually use psychedelic mushrooms, visionary cacti, plant-based snuffs, and psychedelic plant mixtures such as ayahuasca, containing two or more plants indigenous to the region of the Amazon. In West Africa, the powerful psychedelic iboga is used for healing. And in the modern secular world, in addition to all the traditional fungal- and plant-based drugs, there are many purified or synthetic chemical psychedelics, including LSD, dimethyltryptamine (DMT), mescaline, psilocybin, and ketamine.

All these drugs are widely used in modern secular societies, especially by young people. The psychedelic revolution in the 1960s soon ran into legal prohibitions, but nevertheless it has grown and flourished underground, especially through clubs, raves and festivals. At least as many people are taking psychedelics today as in the 1960s.[5]

In spite of their illegality, these substances have even entered the mainstream business world, albeit in an attenuated form. In 2017, *The Economist* ran a feature article entitled 'Turn on, tune in, drop by the office' on technical professionals and others in Silicon Valley who try to get an extra competitive edge by taking microdoses of LSD.[6] Meanwhile, since around 2010, there has been a resurgence of legal research into the effects of psychedelics, as discussed later in this chapter. A best-selling book by Michael Pollan, *How to Change Your Mind: The New Science of Psychedelics*,[7] published in 2018, was a landmark in popularising this new research and in bringing the discussion of psychedelics into the mainstream.

Many psychedelic experiences appear to connect with a non-human world, a realm of nature spirits, demons, angels, gods, goddesses and God. Some researchers have proposed that psychedelics be renamed *entheogens*, a term that literally means 'generating God within'. Some people call them *hallucinogens*, emphasising their role in inducing hallucinations. But this is a narrow definition, because hallucinations are not their only effects; mescaline, for example, leads to a striking enhancement of colour perception. I prefer the more usual term *psychedelic*, which means 'psyche-revealing'.

The psychiatrist Humphry Osmond coined this word in 1956 in his correspondence with the British writer Aldous Huxley, who had written a book, *The Doors of Perception* (1954), describing his very positive experiences taking mescaline. Huxley suggested that these mind-altering drugs be called *phanerothymes*, from the Greek words for 'manifest', *thanenros*, and 'spirit', *thymos*.

In a letter to Osmond, he wrote:

To make the mundane world sublime
Take half a gram of phanerothyme.

Osmond responded:

To fathom Hell or soar angelic
Just take a pinch of psychedelic.

Osmond's suggestion has prevailed.

Everyone agrees that psychedelic experiences are associated with physical and chemical changes in the brain, which can be studied by brain scanning and other medical technologies. But are the experiences themselves nothing but changes in brains? Or are the changes in brains related to mental and spiritual connections that go beyond the brain? Can people really contact other forms of consciousness beyond the human level, as many seem to?

At the end of this chapter, I return to the interpretation of psychedelic experiences. But mostly I will discuss the experiences themselves and scientific research on them. I start with cannabis, and then discuss several psychedelics. I will not attempt to consider the whole range of psychoactive substances, including coffee, tobacco, opioids, cocaine, painkillers, antidepressants and other psychiatric prescription drugs.

Cannabis

In 1968, I received a Royal Society grant to work in the Botany Department of the University of Malaya on rainforest ferns. On my way to Malaysia, I spent six weeks travelling through India. I had intended to stay for a week, but soon after I arrived in New Delhi, I ran into a friend from Cambridge, Johnny Parry, who was doing fieldwork for his anthropology PhD in a remote village in the Himalayan foothills. We met by chance, and he invited me to go back with him to his village; I went.

It was several hours' walk from the nearest roadhead, which we reached after a long journey by train and bus. Johnny was living in the house of a Brahmin family, who were herbal doctors or *vaids*. There was no electricity, no radio, and no television. Everyone slept on mats on the cow-dung plastered floor. Johnny spoke the local language, Pahari, and as we walked through the village in the following days, we chatted with farmers, people in teashops and local artisans. He had been living in the village for more than a year and knew most of them well.

One day we went for a walk along a valley leading into the mountains. Beside a fast-flowing river, we saw a cave, in whose entrance an orange-robed figure was sitting. Johnny explained that he was the local sadhu. The sadhu called out an invitation to visit him. As we sat with him in the entrance of the cave, he produced a *chillum*, a conical clay pipe, and packed material into it, lit it and invited me to have a smoke. Before lighting it, he invoked the god Shiva, and told me that we were smoking Shiva's holy plant. He showed me how to smoke the chillum and I took several puffs.

This was the first time I had taken cannabis. The world was transformed. When I went outside the cave into the sun, as I stood on the grass and looked up at the mountains, I felt a blissful sense of connection with a conscious presence far greater than my own. It was one of the happiest and most exhilarating moments of my life. It left me in no doubt that cannabis could open a door to spiritual experiences. The fact that so many sadhus in India smoke cannabis as part of their spiritual practice shows that this is a widespread belief.

The smoking of cannabis as a spiritual practice is also widespread among some Sufis, members of mystical groups within Islam. When I was working in India in the 1970s, I spent several weeks in the Kashmir Valley, where I was carrying out field experiments with chickpeas. In Srinagar, I visited the shrine of HazratBal, a Muslim holy place believed to contain a hair of the prophet Mohammed. I was surprised to find that it contained a hall in which *fakirs*, or Muslim holy men, assembled to smoke hashish.

Among Muslims who take cannabis, it is often believed that it carries the spirit of the mysterious figure of Khidr, the 'green one', an immortal prophet, often identified with the ascended prophet Elijah, who did not die, but was carried up to heaven in a whirl-wind (2 Kings 2: 11). There are many legends about Khidr, just as there are about Elijah, and Khidr is variously portrayed as a protector, trickster, mystic and saint,[8] and sometimes called the patron saint of cannabis.[9]

A report in the *Guardian* newspaper in 2009 described traditional Sufi celebrations at the shrine of Shah Jamal in Lahore, Pakistan, accompanied by the consumption of cannabis:

> Dance is a popular spiritual expression at shrines such as Shah Jamal. Many aspiring fakirs, aided by the hypnotic beats, dance to find a centre within their bodies and an opportunity to connect with the centre of the universe. The symbol of the lover dancing ecstatically in the presence of the beloved expresses musical and bodily harmony. Physical or emotional intoxication goes hand in hand with the idea of drowning in music, recalling the relation between spiritual ecstasy and intoxication in Sufi culture and poetry . . . Social anthropologist Lukas Werth recalls one of the adherents claiming that 'charas [cannabis] is a bus driver to God.' In this sense, Lukas suggests, the intoxicant is seen as a 'method to open the mind for the divine.'[10]

Some Muslims strongly disapprove. They are against ecstatic practices and the taking of cannabis, and in recent years suicide bombers have attacked Sufi shrines while devotees were celebrating their rituals. This conflict between mystical and fundamentalist Muslims is not new. The popularity of Sufi shrines was responsible for the rise of the Wahhabist movement, now predominant in Saudi Arabia, which first arose as a reaction against the Sufi shrine culture in the Arabian peninsula in the eighteenth century.

Within traditional Islamic societies, the explicit Koranic

prohibition against alcohol was often extended to cannabis as well, although it was not mentioned in the Koran. But on the whole, cannabis was tolerated. Its use was already widespread in the Middle East and Asia before the rise of Islam.

By contrast, within traditional Christian societies, the use of alcohol was generally accepted, even within monasteries, some of which made their living by brewing beer. Cannabis was much less widely known and used. However, it took on a central role in the Bible-based religion of Rastafarianism, which arose in Jamaica in the 1930s. Rastafarian communal meetings involve singing, chanting and the smoking of cannabis, for which they often use the Indian name *ganja*; it is regarded as a sacrament and referred to as 'the holy herb'.[11] Some Rastafarians believe that the Tree of Life in the Garden of Eden (Genesis 2: 9) was in fact the cannabis plant, and they interpret several other Bible passages as promoting its use.[12]

In Europe, cannabis was cultivated primarily for the production of hemp fibre and the making of rope, but was widely known for its medicinal properties and employed as a herbal remedy for a wide range of complaints for many centuries.[13] Queen Victoria took tincture of cannabis to relieve menstrual pains. In Britain, it was banned for general use only in the 1920s, and banned for medical use only since the 1970s.

Under the British Raj in India, the sale of hashish was a government monopoly. I learned this to my surprise in the 1960s, when I was dining on the High Table of a Cambridge college. I was sitting next to the Bursar, who told me that as a young man he had been in the Indian Civil Service, where his last post had been as Government Commissioner for Opium and Hashish in the United Provinces. He had been a drug dealer on a vast scale.

This government monopoly continued after Independence in India and Pakistan. When I was travelling in Pakistan in the 1970s, I noticed a shop in the centre of Lahore called 'Government Opium

Shop'. I went in to have a look, and saw large brown blocks of hashish on sale, stamped 'Government of Pakistan'.

Attitudes to cannabis are rapidly changing, and as a result of the reform of drug laws, cannabis is legally available for medical use and for recreational purposes in a range of states in the US and in several other countries, including Canada. This legalisation is also facilitating scientific research on the effects of this drug.

Cannabis plants contain up to 100 different cannabinoid compounds, of which the principal psychoactive compound is delta-9-tetrahydrocannabinol (THC), acting on the central nervous system primarily via cannabinoid 1 (CB_1) receptors, which are distributed throughout the brain, especially in the frontal regions.[14]

Another kind of cannabinoid receptor molecule, CB_2 is found mainly in immune cells and in the gut. These receptors are evolutionarily ancient, and are found not only in mammals, but also in birds, reptiles, fish, and in many invertebrates, including crustaceans, like lobsters, and worms.[15]

These various receptor molecules respond to compounds produced within the body, called endocannabinoids. The primary roles of this system seem to be in modulating recovery from stress, and in the modulation of immune and inflammatory responses, as well as heart rate, blood pressure, and breathing.[16] In mammals, one of these cannabinoid molecules, produced within the nerve cells, is called anandamide, from the Indian word *ananda*, meaning 'bliss'.

This system of modulating chemicals and their receptors is called the endocannabinoid system. In mammals, this system plays an important role in the control of mood, appetite, pain-sensation and memory. In women, the highest anandamide levels in the blood occur around the time of ovulation.

Another important component of cannabis is cannabidiol (CBD), which does not give a 'high', but is responsible for some of cannabis's medical effects.[17] Cannabinoids can affect almost every organ in the body. The many tissues with cannabinoid receptors and the large number of cannabinoid compounds mean that the psychological and medical effects of cannabis depend on complex interactions.[18]

In high doses, especially if eaten, cannabis can induce hallucinations, but for most people its euphoric effects are often combined with a greater ability to concentrate on sensory inputs: to listen more fully to music, or to look more attentively at a plant, or to engage more deeply in conversations.

But the effects vary – sometimes instead of promoting engagement with other people, cannabis can induce a sense of separation and detachment. Some people become anxious or paranoid, and sometimes cannabis triggers psychosis. It can also be habit-forming and lead to social isolation and a decrease in motivation. Some regular cannabis users feel trapped in a pattern of dependence from which they would rather be free.[19]

The legalisation of cannabis in some US states and elsewhere has meant that there are now attempts to categorise experiences scientifically, if only so that different strains of cannabis can be described more accurately to potential purchasers. Through selective breeding, there are now strains that are exceptionally high in THC and low in CBD, and vice versa.

These have different effects: THC binds to the CB_1 receptors in the brain, activating them, and to CB_2 receptors elsewhere in the body, whereas CBD inhibits these receptors. In combination, these two components of cannabis produce a subjectively more mellow effect than THC alone, and CBD reduces the incidence of paranoia.[20]

Most people do not drink alcohol for spiritual reasons, except in religious rituals like Passover and Holy Communion, and most people do not take cannabis for spiritual reasons either. They take it because they enjoy it and it makes them feel happy. Consuming cannabis is not a spiritual practice for most people most of the time. But it can facilitate spiritual experiences through helping people to be in the here and now.

In those parts of the United States where cannabis has been legalised, there is a serious debate among Christians and Jews about its spiritual value. Some argue that it is forbidden by biblical injunctions against drunkenness. Others point out that cannabis is nowhere mentioned in the Bible, and that it cannot be equated to alcohol. In any case, in the Bible the use of alcohol is accepted, even welcomed, as in Psalm 104: God has given 'wine to gladden the heart of man.' Only the excessive consumption of alcohol is condemned.[21]

And in the New Testament, Jesus was certainly not against wine. His first miracle was the turning of water into wine at a wedding feast after the initial supply had run out, so that people could drink even more (John 2: 1–11). He was not preaching moderation; he was encouraging celebration. And countless Christians commemorate him ritually through drinking wine in Holy Communion.

Many Jewish rabbis, Christian priests and ministers now accept the use of cannabis for medical purposes, and some rabbis have opened cannabis dispensaries themselves.[22] Among evangelical Christians, there are now organised groups where people take cannabis together in the context of Bible study groups. One participant said, 'I believe that consuming cannabis brings me closer to Jesus. It gives me that sense of awe, the spiritual experience I was always looking for in church.'[23]

Some Jewish people have now incorporated the taking of cannabis into their Havdalah gatherings. These traditional ceremonies mark the end of the Sabbath or other holidays, and the beginning of the mundane period that follows. They involve the blessing of an overflowing cup of wine, lighting a candle and smelling sweet spices, together with prayers and joyful quotations from the Hebrew Bible. In some gatherings, the participants now take cannabis together as part of the ceremony, including cannabis-infused bagels, and find that the cannabis enhances their experience of this sacred time.[24]

Meanwhile, for some people, cannabis facilitates the experience of sacred places. In Britain, there is a small underground movement called 'Cathedrals on Cannabis', made up of people who find that taking cannabis before visiting a cathedral can intensify the experience and give a greater sense of holiness and divine presence. Some people say a grace over the cannabis before taking it, and ask for their experience to be blessed. Cathedrals are, after all, buildings designed to have powerful inspirational effects on our consciousness through their sacred geometry, vast proportions, stained-glass windows and echoing acoustics, and cannabis can intensify these effects.

Many people take cannabis outdoors in gardens, parks, forests, and sacred groves and find that it can enhance their sense of connection with the more-than-human world.

The chemical contexts of psychedelics

Just as cannabinoid drugs work on the nervous system by binding to receptors that are naturally present as part of the endocannabinoid system, other psychoactive drugs affect the activity of nerve cells by acting on other chemical receptor systems.

Our nervous systems work both electrically and chemically. Nerve cells, or neurons, meet each other at junctions called synapses, where the release of 'chemical messengers', called neurotransmitters, affects the electrical activity of the adjacent neuron. Some neurotransmitters make the adjacent cells more likely to fire an electrical impulse, while others inhibit this response and make an impulse less likely.

In brains there are many different neurotransmitters. Most of them are fairly simple molecules. Some are amino acids (which are the building blocks of proteins) like glutamate, aspartate and glycine. Another is an amino acid that is not used in proteins, gammaaminobutyric acid (GABA), as discussed in Chapter Three in the context of fasting, which increases the levels of GABA in the brain. Others, like dopamine, noradrenaline and serotonin, are closely related to amino acids. Most psychoactive drugs exert their effects by altering the neurotransmitter effects of the dopamine and serotonin systems. They seem to do so because the chemical structures of these drugs are similar to the natural neurotransmitters.

The phenethylamine family of substances

Dopamine belongs to a family of chemicals known as phenethylamines. Several members of this chemical family are important physiologically or act as drugs. They are described in more detail in Figure 2. Two of the best known are adrenaline and noradrenaline, which are called epinephrine and norepinephrine in the US. Both modulate the activity of the sympathetic nervous system, in which noradrenaline is a neurotransmitter and adrenaline a hormonal stimulant. The secretion of adrenaline from the adrenal glands stimulates the 'fight-or-flight' response and its associated emotions. Adrenaline is also used medically to stimulate the heart in emergencies like cardiac arrest and anaphylactic shock.

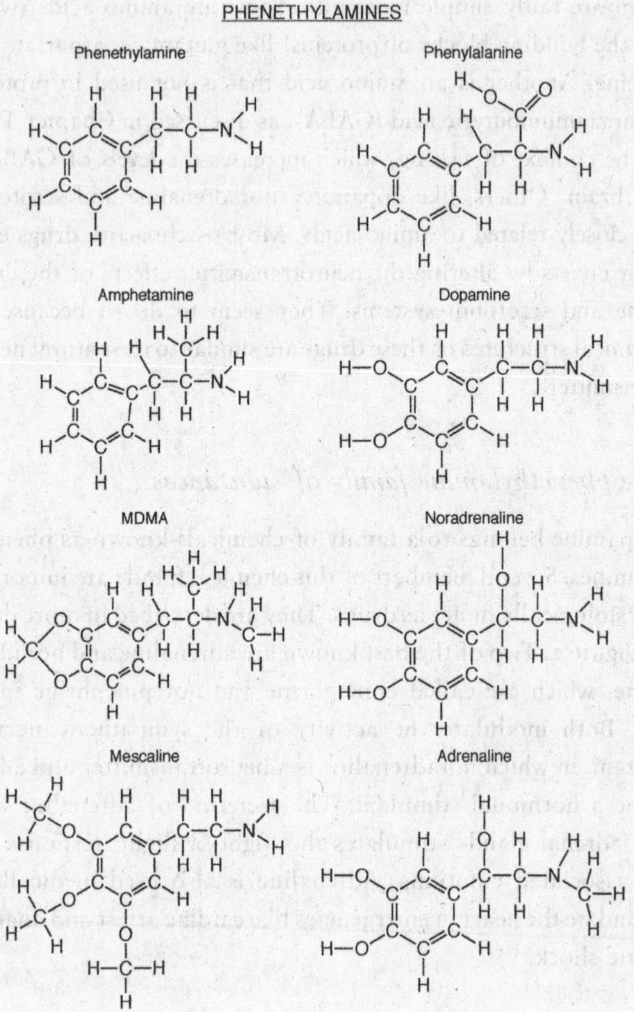

PHENETHYLAMINES

Figure 2. Phenethylamines.
Phenethylamines are closely related to the amino acid Phenylalanine (top right), one of the 20 amino acid building blocks of proteins. Phenethylamine is the molecular structure on which a large family of hormones, neurotransmitters and psychoactive drugs are based. In the right hand column are the structures of the neurotransmitters Dopamine (short for Dihydroxyphenethylamine) and Noradrenaline (Norepinephrine), and the hormone Adrenaline (Epinephrine); in the left hand column, the drugs Amphetamine, MDMA (Methylene dioxy methamphetamine) and Mescaline.

Human-made drugs in this phenethylamine family include amphetamine and methamphetamine (illegally sold as 'crystal meth'), both powerful stimulants, and seriously addictive. MDMA or Ecstasy is also a kind of amphetamine (the A in MDMA stands for amphetamine, as in 3,4-MethyleneDioxyMethAmphetamine), widely used for its effects in inducing euphoria, increasing empathy, heightening sensations and enhancing the pleasure of dancing. The most notable psychedelic in this chemical family is mescaline (Figure 2), which occurs naturally in the peyote cactus, the San Pedro cactus and in a few other plant species, including *Acacia berlandieri*.

The American chemist Alexander ('Sasha') Shulgin (1925–2014), together with his wife Ann, made a systematic exploration of the psychoactivity of more than 170 different phenethylamine molecules. He was the ultimate drug designer. When he synthesised new drugs, he could make an informed guess about their effects, but it was only a guess. He did not know. The only way to find out was to make the compounds and try them himself:

> Even when the compound emerges as a new substance, tangible, palpable, weighable, it is still a tabula rasa in the pharmacological sense, in that nothing is known, nothing can be known, about its action in man since it has never been in man. It is only with the development of a relationship between the thing tested and the tester himself that this aspect of character will emerge and the tester is himself as much a contributor to the final definition of the drug's action as is the drug itself. [25]

Some of the chemicals had no positive effects, and some seemed toxic. For those that seemed harmless and interesting, Shulgin moved to the second stage of testing with his wife. After they had established that a drug was worth exploring further and had worked out appropriate dosages, their volunteer research panel then tried it. Together they came to a consensus view on the drug's effects.

I had the good fortune to know Sasha and Ann Shulgin, and in the early 1980s I attended several symposia at the Esalen Institute, at Big Sur, on the coast of California, where they shared some of their recent discoveries. Together they wrote a book in which they documented their many years of research. Part of this book, published in 1991, is like a cookbook, describing how to synthesise the various chemicals, and – rather like a field guide to fungi – with tasting notes and warnings. The book is called *PIHKAL*, which stands for 'Phenethylamines I Have Known And Loved'.[26]

Sasha Shulgin devoted his life to this research, because he thought that psychedelic drugs were 'treasures'. He enjoyed them for their effects on heightening the perception of colours, or enhancing the senses of touch, smell and taste, or 'the deepening of emotional rapport with another person, which can become an exquisitely beautiful experience, with eroticism of sublime intensity.' But above all, he wrote, 'I deem myself blessed in that I have experienced, however briefly, the existence of God. I have felt a sacred oneness with creation and its Creator.'[27]

The Shulgins' most influential research was on MDMA, which Shulgin first tested on himself in 1976. The chemical itself had originally been made and patented by the German drug company Merck in 1912, but there were no records of any tests on humans. Shulgin found its effects remarkable. After taking a 120 mg dose, he wrote in his notebook, 'I feel absolutely clean inside, and there is nothing but pure euphoria. I have never felt so great, or believed this to be possible. The cleanliness, clarity, and marvellous feeling of solid inner strength continue throughout the rest of the day, and evening, and through the next day . . . All the next day I felt like "a citizen of the universe" rather than a citizen of the planet.'[28]

When the Shulgins and their research group first explored the effects of MDMA, it was not illegal, because it was almost unknown. It soon became clear that it had great potential for use in psychotherapy, enabling patients to explore traumatic events in their life, with the help of their therapists, or painful aspects of relationships, without the usual fear and resistance.

When I myself first took MDMA in California in the early 1980s, I did so with a small group of friends in a beautiful woodland environment, and the effects were blissful. We imagined that this benign compound could change the world through the healing effects of love. At that time, it was usually called 'Adam'. One psychotherapist described it as 'penicillin for the soul'.[29]

Within a few years, MDMA escaped from the world of psychotherapists and researchers and became immensely popular in parties and clubs. It helped give rise to the new rave culture, which had large-scale euphoric effects and a huge cultural impact. But there were some casualties. Scary stories started appearing in the media, and the drug was made illegal in the US in 1985, and official research on its therapeutic effects ground to a halt. Only in 2016 was a new research project on the uses of MDMA in psychotherapy officially approved in the US, particularly for the treatment of post-traumatic stress disorder. In 2017, a project began in the UK for research on its use in the treatment of alcohol addiction.

The tryptamine family of substances

Another chemical family, the tryptamines, affect a different neurotransmission system, based on serotonin. Serotonin is 5-hydroxy tryptamine (see Figure 3 for chemical details). Other members of this family include the hormone melatonin and the psychedelic psilocybin, found in more than 200 species of 'magic mushroom', including *Psilocybe semilanceolata*, the Liberty Cap, which grows in moist grasslands throughout Europe, and *Psilocybe cubensis*, native to Central and South America, the most commonly cultivated and (illegally) available species.

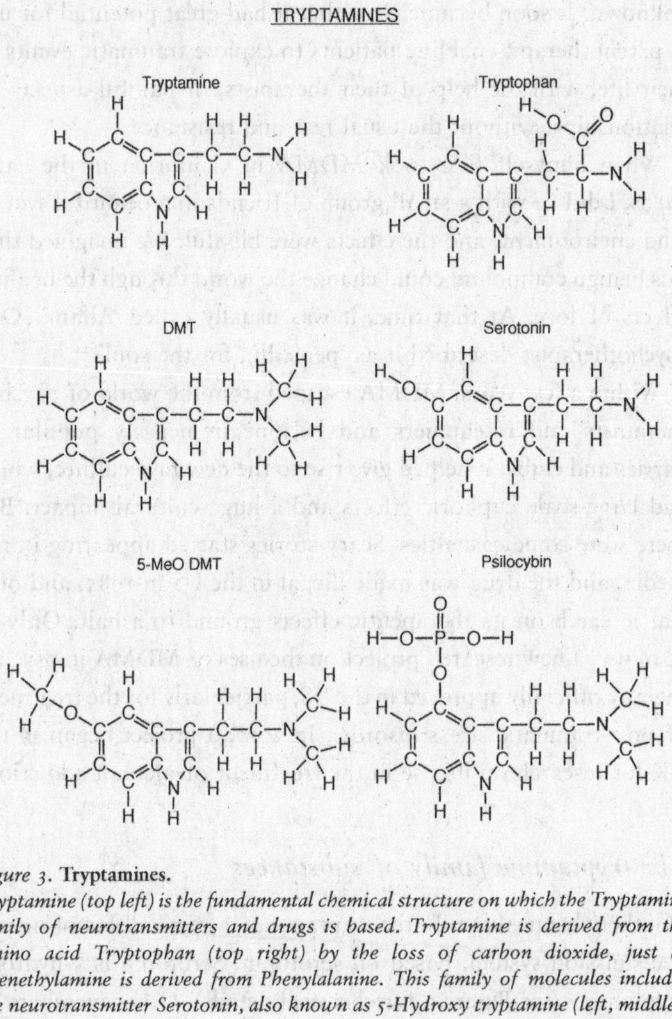

TRYPTAMINES

Figure 3. **Tryptamines.**
Tryptamine (top left) is the fundamental chemical structure on which the Tryptamine family of neurotransmitters and drugs is based. Tryptamine is derived from the amino acid Tryptophan (top right) by the loss of carbon dioxide, just as Phenethylamine is derived from Phenylalanine. This family of molecules includes the neurotransmitter Serotonin, also known as 5-Hydroxy tryptamine (left, middle), and the psychoactive compounds Psilocybin, Dimethyltryptamine (DMT) and 5-Methoxy dimethyl tryptamine (5-MeO DMT).

The intense and fast-acting psychedelic dimethyltryptamine (DMT), which occurs naturally in a range of plant species, is also closely related in structure to serotonin, as is 5-methoxy DMT

(5-MeO DMT), which is highly psychoactive and occurs naturally in several plant species, and also in the skin of the Colorado River toad.

The chemist Albert Hofman first created the synthetic molecule LSD in Switzerland in 1938, and first experienced its effects accidentally in 1943, when he absorbed some of it through his skin when working in the lab. As he bicycled home, he felt dizzy, and when he arrived he lay down: 'In a dreamlike state, with eyes closed (I found the daylight to be unpleasantly glaring), I perceived an uninterrupted stream of fantastic pictures, extraordinary shapes with intense, kaleidoscopic play of colours.'[30]

LSD is not chemically a tryptamine, but much of its effect seems to depend on its ability to bind to serotonin receptors. Its full chemical name is lysergic acid di-ethyl amide; it is closely related to ergine (lysergic acid amide), a compound found in ergot, a fungal disease of rye and barley, and also in the seeds of morning glory plants, and other plant species.[31]

Some scholars have suggested that ergot was a component of the visionary brew given to initiates in the cave of Eleusis in ancient Greece, as part of the celebration of the Eleusinian mysteries. The ritual journey of the Eleusinian devotees into the cave, their experiences within it, including the effects of the sacred brew, and their return to the light of day, re-enacted the myth of the abduction of the vegetation goddess Persephone from daylight on earth to the darkness underground by Hades, the king of the underworld. Through the intervention of her mother, the harvest goddess Demeter, and her father, the sky god Zeus, Hades released her and she returned to the light.[32]

In 1997, Sasha and Ann Shulgin published a sequel to *PIHKAL* called *TIHKAL*, standing for 'Tryptamines I Have Known And Loved', in which they explored more than fifty tryptamine compounds.[33] None were as interesting as the naturally occurring DMT, 5-methoxy DMT and psilocybin. Some had unsettling and bizarre effects, like DIPT (di-isopropyl tryptamine). Shulgin noted that the most striking effect of this compound was that it affected hearing rather than seeing, unlike most psychedelics. All frequencies seemed to shift to a lower pitch:

> All familiar sounds became foreign, including the chewing of food . . . Music was rendered completely disharmonious although single tones sounded normal. There were no changes in vision, taste, smell, appetite, vital signs, or motor coordination. [But with a higher dose] There was a feeling of foreboding . . . The voices of people were extremely distorted – males sounded like frogs – children sounded like they were talking through synthesisers to imitate out-of-space people in science fiction movies. In fact, I felt that I was somehow sent into an anti-universe, where everything looked the same as normal, but was a cold and empty imitation. I felt I was a fallen angel.[34]

The fact that the auditory effects of this drug were so unusual helps draw attention to the fact that psychedelics primarily affect seeing. Sometimes they produce synaesthesia, where one sense becomes another. Music often takes on visual forms. But synaesthesia rarely happens the other way round; external visual images do not usually turn into sounds.[35]

Some chemical molecules may be more likely to create positive, connected, happy, flowing experiences; others may be more likely to produce negative experiences, as DIPT did for Sasha Shulgin. But positive and negative effects, good trips and bad trips, are not only a matter of chemistry, but also depend on the way a person responds to the drug. Early researchers on psychedelics in the 1950s and 1960s soon realised that a drug's effect depended on three factors: the drug itself; the setting in which it was taken; and the 'set' or psychological situation of the person who took it.[36]

The traditional use of psychoactive drugs in indigenous

ceremonies is not primarily about trips, but about initiation, healing and community decision-making. But here too the context, the setting and the set, are of vital importance; the healing effects of psychedelic ceremonies are not purely a chemical effect of the medicine.

Psychedelic seeing

One of the most influential writers on psychedelic seeing was Aldous Huxley (1894–1963). His book, *The Doors of Perception*, borrowed its title from the poet William Blake: 'If the doors of perception were cleansed everything would appear to man as it is, infinite.' In particular, the doors of vision were cleansed. As Huxley put it, 'Visual impressions are greatly intensified and the eye recovers some of the perceptual innocence of childhood.'[37]

In his follow-up book, *Heaven and Hell* (1956), Huxley extended his survey to include LSD:

> The typical mescaline or lysergic acid experience begins with perceptions of coloured, moving, living geometrical forms. In time, pure geometry becomes concrete, and the visionary perceives, not patterns, but patterned things such as carpets, carvings, mosaics. These give place to vast and complicated buildings, in the midst of landscapes, which change continuously, passing from richness to more intensely coloured richness, from grandeur to deepening grandeur. Heroic figures, of the kind that Blake called 'The Seraphim,' may make their appearance, alone or in multitudes. Fabulous animals move across the scene. Everything is novel and amazing.[38]

Huxley saw these experiences as visits to what he called 'the antipodes of the mind', and described these other worlds as 'more or less completely free of language, outside the system of conceptual thought.' He thought that the landscapes of the mind's antipodes are not merely personal, but exist in the collective unconscious.[39] He was struck by the close similarity between these visionary experiences and the fairylands of folklore and the heavens of religion.

In ancient Greece and Rome, there were stories of the garden of the Hesperides, the Islands of the Blest, and other luminous places. In the Hindu epic, the *Ramayana*, the land of Uttarakuru 'is watered by lakes with golden lotuses. There are rivers by thousands, full of leaves the colour of sapphire and lapis lazuli; and the lakes, resplendent like the morning sun, are adorned by golden beds of red lotus.'[40]

Similar gem-filled landscapes appear in the Bible, as in the prophet Ezekiel's version of the Garden of Eden (Ezekiel 28: 13) and in the vision of the New Jerusalem at the end of the Bible, in the Book of Revelation: 'The wall is built of jasper, while the city is pure gold, clear as glass. The foundations of the wall of the city are adorned with every jewel.' (Revelation 21: 18–19.)[41]

Huxley argued that in human cultures the causal chain began with visionary experiences of other worlds, which were then made visible on earth through gems, polished surfaces, beautiful gardens and coloured glass, as in the great medieval cathedrals, with their rose windows like vast magical flowers. And then from the earth, minds mounted again to the other world, like Plato's ideal world of Forms, above and beyond the world of matter. In the words of Socrates, in the dialogue *Phaedo*, 'In this other earth the colours are much purer and much more brilliant than they are down here . . . the very mountains, the very stones have a richer gloss, a lovelier transparency and intensity of hue . . . The view of that world is a vision of blessed beholders.' In this realm of Forms or Ideas, to see things 'as they are in themselves' is to be in a state of inexpressible bliss.[42]

For some people, the visionary experience starts from light itself.

One woman described to me an LSD trip in which she found herself travelling first through vast regions of desert and then beginning an upward journey into the light: 'As I entered more and more it seemed to penetrate me, through every cell. I was aware of it entering every part of me. The feeling of being surrounded and penetrated by Light lasted what appeared to be a long, wonderfully long time . . . I became one with the Light. I was the Light.'

And then forms appeared: 'I saw the most intricate beautiful meshing of what you might describe as gears in geometric designs so complex and complicated that I was amazed and awed . . . Then I was in an unearthly world of more and more and more pure, pure beauty. The colours and forms were again unknown, the beauty excruciating.'[43]

Psychedelics are not the only gateways to these visionary realms. Some people have visions spontaneously. Others have them after spiritual practices like fasting, or spending long periods in darkness. Some people have very vivid visual imaginations. And all of us have dreams, usually several every night, although we normally forget them. In our dreams we find ourselves in self-luminous worlds that we do not consciously create, and some people have very intense dreams. Psychedelic visions can be even more intense, with far brighter colours.

For some visual artists, psychedelic visions are inspirations for their work. And, historically, they provided a powerful stimulus and motivation for many people in the field of computer graphics. In 1996, a series of reports in the US media about the use of psychedelics in Silicon Valley suggested that these experiences played a major role in stimulating creativity. Timothy Leary was quoted as saying, 'The Japanese go to Burma for their teak, and they go to California for novelty and creativity. Everybody knows that California has this resource thanks to psychedelics.'[44]

A sceptical journalist who did not believe this claim set out to prove it was false. She went to the SIGGRAPH convention, the largest gathering of computer graphics professionals in the world, and interviewed a sample of 180 delegates, asking them, 'Do you take psychedelics and is this important in your work?' To both questions, all 180 answered 'yes'.[45]

Effects of psychedelics on brains

Research on the effects of psychedelics on brain activity using modern brain-scanning techniques began around 2010. There have now been many published studies, mainly on the effects of psilocybin and LSD. Both drugs reduce the amount of blood flow and activity in the default mode network (DMN), an interconnected system of brain regions involved in rumination and internal dialogue, or 'mental chatter'.[46]

The activity of the DMN is also decreased by meditation (*SSP*, chapter 1) and outward directed activities, including sports. This measurable reduction in the activity of the DMN is not surprising. People taking these intense psychedelics do not find themselves engaged in internal dialogues and other activities associated with the activity of the DMN. They are immersed in their visionary experience, very much in the present.

In studies of people who were having intense visual experiences on LSD, the visual parts of their brain were much more active than in control subjects who have been given a placebo instead of the LSD.[47] The blood flow in the visual regions increased, and correlated well with self-reported hallucinations. These studies also showed that the visual cortex was in communication with a wider range of other brain regions under the influence of LSD than in subjects who had been given a placebo; there was more interconnection. In other words, many brain regions not normally involved in vision contribute to visual processing, which may help to explain why people tend to experience dreamlike hallucinations.

Further studies that looked in more detail at the visual regions of the brain found that the activity of the visual cortex was much better coordinated in subjects who had received LSD than the placebo. All had their eyes closed, but with LSD, brain activity took place as if the eyes were open, and as if they were looking at objects that were spatially localised.[48]

When the people who had these experiences were interviewed two weeks after their trial sessions, their ratings for the personality trait 'openness', linked to imagination and creativity, were higher than in the control group who had taken the placebo. The researchers suggested that 'psychedelics may serve as a kind of "existential shock" therapy, where the profound psychological experience can lead to a change in behaviour and outlook.'[49]

In another study, volunteers were tested to explore the effect of LSD on their responses to music. Volunteers had their brains scanned while they were on LSD, or on a placebo, and when they were listening to music, and not listening to music. In those who were listening to music on LSD, there was an increased connectivity between a part of the brain called the parahippocampal cortex (PHC) and other parts of the brain, especially at the visual cortex. The PHC is connected with emotion, memory and 'ego functions'.

The researchers established that the direction of this connectivity was from the PHC to the visual cortex, not vice versa. This effect was correlated with self-reports of visions (with eyes closed) of scenes from the person's past. Music and LSD together increased autobiographical mental imagery. The more intense the overall LSD experience, the more emotionally arousing the music felt, the strongest emotion being 'transcendence'.[50]

Such studies have only just begun. So have studies on the effects of psychedelics on nerve cells grown in the laboratory, where it turns out that a range of psychedelic compounds, including LSD and DMT, promote the growth and branching of neurons.[51] We can expect to learn many more details about the effects of psychedelics on brain activity and on nerve cells. But these studies reveal very little about the experiences themselves. In fact, they bring us up against the 'hard problem' in the philosophy of mind. The 'hard problem' is the very fact of human consciousness. Objective brain measurements tell us about activities in brains, but not about people's experiences. The subjects have to describe their experiences themselves. They are subjective, not objective.

These kinds of studies show that there is a relationship between these experiences and brains, but do not prove that psychedelic experiences are nothing but brain activity. When someone speaks to you on your mobile phone, you experience her voice and react to what she is saying. But this does not prove that the person speaking to you and the meaning of what she is saying are nothing but the measurable physical activity inside the phone or the vibratory physical changes inside your ears and brain. Your recognition of her voice, your understanding of her meaning, your picking up nuances from her intonations – these are conscious experiences, and not just physical vibrations.

Interpretations of the effects of psychedelics depend on our worldviews. Materialists believe that matter is the only reality, and minds are nothing but the electrical and chemical activity of brains. Psychedelic experiences are nothing but chemical disturbances of brain activity. These experiences may make people feel they are travelling out of their body, or that they are encountering spiritual beings, or that they are in the presence of God. But all these feelings and experiences are in fact generated inside their heads. They are not in contact with other realms of consciousness 'out there'. These experiences are produced inside their brains, and insulated within their skulls. They are private.

Panpsychists or animists believe that there are many forms of mind or consciousness in systems at all levels of complexity. In psychedelic experiences, human minds may come into relationship with some of the many other minds in nature, including animal minds, and perhaps plant minds, and also minds far more extensive than our own, like that of the sun, or the galaxy, or the cosmos. Minds are not just in brains. In psychedelic experiences, we interact not only with other human minds, past and present, but with non-human minds as well.

Many religious people agree with panpsychists that there are many forms of consciousness in the cosmos, but they also believe in an ultimate form of consciousness that transcends the cosmos, giving rise to it, and sustaining it, and underlying all the minds within it. From this point of view, mystical experiences of an all-sustaining conscious being are not illusions, but direct experiences of reality.

Experimental mysticism

Many people experiencing LSD, magic mushrooms, mescaline and other psychedelics experience a sense of a greater presence, or 'is-ness' or 'beingness'. Aldous Huxley wrote with particular clarity about such experiences. When looking at some flowers after taking mescaline:

> I continued to look at the flowers, and in their living light I seemed to detect the qualitative equivalent of breathing – but of a breathing without returns to a starting-point, with no recurrent ebbs but only a repeated flow from beauty to heightened beauty, from deeper to ever deeper meaning. Words like grace and transfiguration came to my mind, and this of course was what, among other things, they stood for . . . The Beatific Vision, Sat-Chit-Ananda, Being-Awareness-Bliss – for the first time I understood, not on the verbal level, not by inchoate hints or at a distance, but precisely and completely what these prodigious syllables referred to.[52]

One of the pioneering researchers on LSD, Oscar Janiger, summarised his own and other people's experiences as follows: 'A new vista opens up all in a moment and while you feast upon it, time stands still. You have a feeling of nowness. There is no past or future. This ecstasy is the only time you are alive. Happiness is not something to be experienced sometime in the future on vacation or after retirement. It is now.'[53]

This psychedelically induced feeling of connection, unity and conscious presence is similar to mystical experiences that arise as a result of meditation, prayer, fasting or other spiritual practices, or that come spontaneously. Two other pioneering psychedelic researchers, William Richards and Walter Pahnke, described experiences with these drugs as 'experimental mysticism'.[54]

Given these experiences, it is not surprising that some people regard the taking of psychedelics as a kind of sacrament. But for others they are primarily recreational, giving a strong dose of intense fun at weekends and in festivals. Before research on LSD became illegal in 1965 in the US and elsewhere, some researchers gave the drug to many volunteers from a variety of backgrounds, providing them with a safe environment and collecting thousands of self-reports.

In a follow-up study forty years later, nearly all the respondents felt that their experiences with LSD had been positive, and for some they were also spiritual or transformative. About a third of the respondents said that they had experienced persisting beneficial changes from their experiences in these early studies. One said that taking LSD was the 'most extraordinary experience' of his entire life. He described it as 'a genuine peak experience.'[55] Another said, 'I was raised without religion and I was not spiritual until I took LSD. I've been spiritual ever since.' Another said that taking LSD had 'eased his fear of death.'[56]

On the other hand, two thirds of those surveyed said that they were not profoundly influenced by their experiences, which they described simply as curious or inexplicable or interesting. Some said they did not feel drawn to take LSD again.

Thus, for some people experiences with psychedelic drugs, even if intensely enjoyable, do not have much lasting effect. But others have mystical experiences that change their lives.

Psychedelic experiences as rites of passage

In traditional cultures around the world, there are a series of rites of passage, including those associated with birth, young people passing from childhood to sexual maturity, marriage, and death (*SSP*, chapter 5).

These rituals often symbolise a process of death and rebirth. The people being initiated die to their old role in society, pass through a dangerous or frightening period with new and sometimes visionary experiences, enter the spirit world, and then – unless they have actually died – return to the world of adult men and women. Some of the rites may have induced actual near-death experiences (NDEs).

Near-death experiences are commoner today than ever before, thanks to coronary resuscitation and modern medicine. Those who have them often experience going out of their body, passing through a dark tunnel, and entering a realm of light and joy, where they encounter loving beings. Then they come back again. These experiences often transform people's lives and reduce their fear of death. They feel they have died and been born again.[57]

I have already touched on the fact that in ancient Greece and Rome, initiation rites typically involved a ceremonial death and rebirth, as in the initiation into the mysteries of the Mithraic cult.[58] In general, in these initiation rites, there was 'the representation of the passion of a divine or semi-divine being, who is attacked or carried off by infernal powers, descends to the realm of the dead, is liberated by the intervention of some higher divinity, and brought back to the region of light in the presence or company of those assisting in the ceremony.'[59]

Similar patterns are found all over the world. In the lower Congo, initiation ceremonies were called *kimbasi*, which means 'resurrection': 'During a dance the neophytes fall dead, and the sorcerer resuscitates them.'[60] In the Arctic, among the Inuit, 'an *angakok* [medicine man] goes through the ceremony of killing the aspirant to magical powers, and his soul then flies off to probe the depths of sky, sea, and earth, and thus learn the secrets of nature. On its return it resuscitates the body, which has been lying stretched on the frozen ground, and the patient then becomes an *angakok* in his turn.'[61]

The initiation ceremony practised by John the Baptist, described in the New Testament, was one of death and rebirth through drowning, as the candidates were totally immersed in the River Jordan. Was this merely symbolic, or did they have an actual NDE? If John held the people he was baptising under the water just long enough, he would have induced an NDE by drowning, as I discuss in *SSP*, chapter 5.

Some psychedelic experiences are like near-death experiences. Stan Grof, a psychiatrist whose early research took place in his native Czechoslovakia during the Communist period, and later in the United States, studied the responses to LSD by more than 2,000 people. One of his most interesting findings was that this drug could 'induce, without any specific programming and guidance, profound death-rebirth experiences, and facilitate spiritual opening.'[62]

Grof thought that some of these experiences reflected a deep archetypal but unconscious memory of the process of being born. In this birth-process model, as the uterine contractions begin, there is a feeling of being trapped in a dark and menacing world, which leads to a struggle as the baby starts its passage along the birth canal. As it passes through this dark tunnel, it can see light at the end and finally emerges into the light. Grof found a similar sequence in many people's experiences of LSD.

Another researcher, Charles Hayes, found a similar pattern:

> One of the most remarkable features of the psychedelic experience I discerned in the course of my research was a cycle of contraction and relaxation, culminating in a kind of oceanic release. The subject begins to feel closed in – as if some ethereal sphincter is cutting him off from the realm of light and life, sometimes seemingly to the point of strangulation or death, only to open up, usually as a result of his conscious surrendering to the experience, to reveal a world that sparkles anew or alludes to the infinite expanse of the All (or the Shining Void, depending on one's terminology.)[63]

On the other hand, Hayes found that a too-sudden opening up could induce an onslaught of panic, a terror of the vastness of the void, or what Aldous Huxley called 'a horror of infinity.'[64]

When I myself first took LSD in Cambridge in 1970, the effects came on far sooner than I had expected, and I found myself trapped in a realm of uncontrollable hallucinations that made me think I had taken a massive overdose, gone mad and that I would never recover. I thought I was in hell, a fear immediately confirmed by the vision of huge neon signs all around me saying, 'Hell'. Then I felt as if I were drowning, deep underwater. But thanks to re-assurance of a friend who was with me, I calmed down, and then found myself rising up through the water, until I burst out through the surface into the light, into a realm of astonishing beauty and joy.

It was as if I had been liberated, born into a new reality. This experience was a spiritual awakening that changed the course of my life. I knew nothing of the research by Stan Grof and others, and had no expectation of anything like this.

Some substances induce NDE-type experiences even more strongly than LSD. Ketamine, a medical and veterinary anaesthetic, has psychedelic effects at lower doses, and often gives an extreme sense of disconnectedness from the outside world, including out-of-the-body experiences.[65] Spiritual experiences with ketamine are common.[66] In a long series of studies, researchers found that the intravenous administration of ketamine could reproduce all the features commonly associated with NDEs, including feelings of peace and contentment, a sense of detachment from the body, entering a transitional world of darkness as if passing through a tunnel, and emerging into bright light.[67]

DMT is one of the most intense of all psychedelics. When smoked or injected, it has an extraordinarily rapid effect, which lasts for only about ten minutes. People on DMT trips often feel as if they have left their body, and some experience going through a kind of tunnel, into a realm of light, colour and form.[68] When I myself first took DMT, around 1983, I found myself in front of a huge chrysanthemum flower, and whooshed through its centre, as if through a tunnel, emerging into a blissful realm of intense light and colour, with ever-changing, petal-like shapes. I then found myself coming back into my body from very far away. I was given the DMT by my friend Terence McKenna, and when I described my experience to him, he said, 'You have been to the flower heaven.' It felt to me as if I had died, travelled to another realm, and then come back to life again in my body. It was a kind of near-death experience.

In a research project at the University of New Mexico, one of the very few investigations of the effects of psychedelics in the 1990s, the psychiatrist Rick Strassman gave DMT to sixty volunteers, and found that some of them had experiences very like NDEs. And, like spontaneous NDEs, some of these experiences had positive effects, reducing the fear of death and increasing appreciation of life.[69]

In a follow-up study, Strassman interviewed volunteers after their DMT experiences to find out about their long-term practical effects.

He was hoping that their lives would have been changed for the better, but was disappointed to find that for most people they had made little difference. DMT was not inherently therapeutic: 'Without a suitable framework – spiritual, psychotherapeutic or otherwise – in which to process their journeys with DMT, their sessions became just another series of intense psychedelic encounters.'[70]

On the other hand, Strassman was struck by the fact that when he studied at a Zen monastery, nearly all the young American monks told him that psychedelic drugs, especially LSD, had first opened the doors to a new reality for them. These drugs had been a kind of rite of passage that had led them to follow the discipline of a communal, meditation-based ascetic life.[71] But they were not interested in trying to combine their Buddhist practice with any further use of psychedelics.[72]

However, some traditional societies and religions have integrated the use of psychedelics into their rituals, and provide a framework of support and interpretation for these substances' roles in healing and initiation.

Psychedelic rituals and religions

In many cultures, psychedelics have been used for millennia in healing rituals and initiation rites. As I already mentioned, a mind-altering brew was an integral part of initiation into the Eleusinian mysteries in ancient Greece. In India, the Rig Veda, the oldest classical Vedic scripture, contains a collection of hymns to *soma*, a brew whose identity is unknown, but whose effects were clearly powerful. Here are some examples:

> The Soma is full of intelligence. It inspires man with enthusiasm.
> It makes the poets sing. We have drunk the Soma: we have come
> to be immortal, we have arrived at the Light, we have reached
> the Gods.[73]

In ancient Persia, Zoroastrian hymns to the sacred plant *haoma* were similar. What were they taking? No one really knows, but among scholars there are many speculations as to the identity of the plants or mushrooms used in these brews.

One of the most widespread traditional psychedelics is *ayahuasca*, used by the indigenous people of the upper Amazon over a vast region including western Brazil, Peru, Colombia and eastern Ecuador. Drinking this brew is at the centre of initiation rites, and is traditionally used for all major decisions of the tribe, including locating game for hunting. It is still widely used by shamans in healing ceremonies.[74]

But as its fame has spread, so has the demand for taking part in ayahuasca ceremonies. Dozens of centres have sprung up in Peru, where this use of the brew is legal, catering to the thousands of people who travel there to take it. The ayahuasca business is booming. Foreigners are buying land and opening centres, and many people are calling themselves shamans, *curanderos* (healers) or *ayahuasqueros*, even if they have little or no connection with traditional practitioners.[75]

Ayahuasca has now spread worldwide. In 2016, according to the *New Yorker* magazine, on any given night in Manhattan there are about a hundred ayahuasca circles. Most of the people leading them are not from traditional tribal backgrounds.[76] And, for a fee, you can now train to be an ayahuasquero in courses advertised online.[77]

Ayahuasca has become a central sacrament in several Christian psychedelic churches in Brazil. One of them, called Santo Daime, started in the jungles of Acre Province by a black rubber tapper called Mestre Irineu (1890–1971), who was raised Roman Catholic and who encountered the Virgin Mary as the Queen of the Forest.[78] He said that she guided him and showed him the rituals that he should follow. The name of the church came from a hymn, *Dai-me força, dai-me amor, dai-me luz* – 'Give me strength, give me love, give me light.'[79]

The Santo Daime church now has many centres in rural and urban Brazil, in the United States, Europe and elsewhere. This church has been legal in Brazil since 1992, and is also legal in some US states and in the Netherlands. Each ritual session is led by a 'godfather' and begins with Christian prayers, including the Our Father and the Hail Mary. Then the participants drink ayahuasca, called *daime*, as a kind of sacrament. The rituals include the singing of hymns and periods of dancing, as well as times for concentration, sitting quietly.

Another legal ayahuasca church in Brazil is called União do Vegetal (Union of the Plants), founded in 1961. Unlike Santo Daime, which grew out of a poor, mixed-race rubber-tapper culture, União do Vegetal is more middle class, with many doctors, lawyers and politicians. It now has offshoots worldwide, including the United States, where a Supreme Court decision in 2006 legalised its practices.[80]

One precedent for this decision was the legalisation of the Native American Church in 1978. This psychedelic religious movement arose in the late nineteenth century in Oklahoma, combining Christian teachings and morality with the use of peyote, a mescaline-containing cactus. It is now the most widespread indigenous religion among Native Americans in the United States, Canada and Mexico.

Missionaries brought the Christian message to psychedelic-using shamanic cultures in the Americas. An unintended consequence was the evolution of psychedelic churches. Together with psychedelic neo-shamanism, these churches are now spreading into North America and Europe in a kind of reverse missionary movement.

Some people speculate that psychedelics may have played a hidden role in the development of Judaism. Benny Shanon, a Professor of Psychology at the Hebrew University of Jerusalem, suggests that Moses, who led the Jewish people out of slavery in

Egypt, may have had experiences similar to those induced by ayahuasca.

Shanon proposed his 'biblical entheogen' hypothesis after taking ayahuasca many times himself in South America and interviewing many other people who have taken it. In his book, *Antipodes of the Mind* (2002), he sketches out a map of ayahuasca visions, based on about 2,500 experiences. He suggests that several key episodes in Moses' life resemble experiences with ayahuasca, including his encounter with a burning bush in the desert, his magical contest with the Pharoah's sorcerers turning rods into serpents, and the visions on Mount Sinai when he received the Ten Commandments. The conventional view, by contrast, is that Moses' experiences were not drug-induced, although some were facilitated by long periods of fasting.

For Shanon, these experiences suggest that Moses was familiar with an ayahuasca-type brew, which requires two plants: one that contains DMT, and the other a plant that contains harmine and harmaline, which stop DMT being destroyed when ingested. In the Amazon, the DMT component of ayahuasca is provided by a shrub called *chacruna* (*Psychotria viridis*) and the second component comes from a vine, *Banisteriopsis caapi*, containing harmine and harmaline.

In the southern parts of the Holy Land and in the Sinai peninsula, where Moses had his visions, a bush called *Peganum harmala*, or *harmal* in Arabic, contains harmaline and harmine, which are named after it. Some scholars identify it with haoma. This plant is also called Syrian rue, and is still widely used as a herb in the Middle East. Meanwhile, several species of acacia tree contain DMT.

In the book of Exodus, immediately after the account of the experiences of God at Mount Sinai, Moses commanded his followers to make a chest to contain the two stone tablets of the Ten Commandments, known as the Ark of the Covenant. This was made out of *shittim* or acacia wood, and the same kind of wood was also used for the planks that supported it and the altar beside it (Exodus 25: 1–29; 26: 36–7; 27: 1, 6).[81] But, of course, the presence of these plants does not prove that Moses and others knew how to combine them together to make a visionary brew, nor that Moses took such a brew.

Shanon is not alone in his speculations about biblical entheogens. Rick Strassman, who carried out the research on DMT discussed above, has suggested that the experiences of some of the Hebrew prophets were very similar to the effects of DMT.[82]

Whatever Moses and the prophets may or may not have ingested, some researchers have suggested that many thousands of years before these events, psychedelic plants and mushrooms played a major role in the evolution of human consciousness, language and art.[83] Certainly many of the images in prehistoric cave painting, some more than 30,000 years old, resemble images seen in visionary states under the influence of psychedelics. These include human figures with animal heads, including lions and bison, and curious stick figures.[84]

It seems likely that the people who painted them were familiar with altered states of consciousness, but these need not necessarily have been psychedelic; people could have entered these visionary states in other ways, including spending long periods in darkness, or through trance-inducing dances. Some may have occurred spontaneously, without drugs or any special practices. They may not have discovered psychedelic plants and mushrooms.

Nevertheless, some prehistoric cultures did make this discovery, although no one knows how long ago. But there is no doubt that psychedelic substances are still used in shamanic and religious rituals in many cultures today, and that their use is traditional.

Seemingly autonomous entities

In the modern secular world, people who take ayahuasca, DMT, magic mushrooms, LSD and other visionary substances are often surprised to encounter 'entities', some of which are friendly, some neutral and some terrifying. In his research on the effects of DMT, Strassman found that many of his volunteers encountered 'beings', 'aliens', 'guides', and 'helpers'. Some of them looked like clowns, insects, spiders, reptiles, cacti and stick figures. And some of these experiences were like alien abduction scenarios: 'Volunteers find themselves on a bed or in a landing bay, research environment or high technology room. The highly intelligent beings of this "other" world are interested in the subject, seemingly ready for his or her arrival and wasting no time in "getting to work" . . . Their "business" appeared to be testing, examining, probing and even modifying the volunteer's mind and body.'[85]

Strassman's subjects were indeed in a research setting. They were, in fact, being probed and modified by intelligent beings who were ready for their arrival for the testing session. They received an intravenous injection of the drug, and had their blood pressure monitored and blood samples taken while they were under the influence of the DMT. Perhaps the highly intelligent aliens who were testing and probing them were transforms of Strassman and his assistants.

Nevertheless, many people who take psychedelics encounter seemingly autonomous beings, even when they are not in research settings. These beings seem to have a life of their own. Some are friendly, some neutral, some threatening and dangerous. People also encounter deceased ancestors, saints, angels, winged creatures, animal spirits, chimeras – part human, part animal – gods and goddesses, the Buddha, the Blessed Virgin Mary, Jesus Christ, and God.

In the worlds of shamanism and religion, the existence of

autonomous spirit beings, such as animal spirits, deceased ancestors, demons, angels, devas, dakinis, gods and goddesses, is taken for granted. People encounter other kinds of being in their dreams and in visionary states. Some Hindus, for example, have dreams of Ganesh, the elephant-headed god, and other Hindu gods and goddesses, and even discuss them on online forums.[86]

Are these beings simply products of the experiencer's mind? As Terence McKenna commented, 'For the psychedelic voyager, the intuition is made of mind but not of my mind.'[87] Maybe they are part of the collective unconscious, a kind of collective memory. If lots of people in India look at images of Ganesh, and dream about Ganesh, he may take on an autonomous life of his own within the collective unconscious.

If he appears in someone's dream he is a product of mind, but not only of the dreamer's mind. His properties may depend on many human minds, and on the stories told about him. But at the same time, he may be more than a collective product of human minds; he may be a channel for a divine consciousness that interacts with human minds through this biologically impossible form.

For those who believe that there may be many kinds of consciousness beyond those in human and animal brains, the question is, where does all this stop? Are there innumerable ancestors out there in the spirit realm, and animal spirits, and the spirits of extinct species, and the spirits of fictional characters, and the spirits created by myths and images, and human representations of gods and goddesses, and human-animal hybrids like centaurs or Ganesh?

And, beyond this collective realm of forms and memories, are there intelligences throughout the universe – minds in planets, stars and galaxies? Is there a cosmic mind? Is there a mind beyond the cosmos? Or is it all in our brains?

In thinking about these questions, it is helpful to start from dreams. We all dream, though many of us forget most of our dreams, and do not take very seriously the ones we remember. They rarely change our lives. But some people pay attention to their dreams, and when they do so, they find that sometimes they dream about things that happen later, usually within a few days. These dreams suggest that we can be linked to our futures.[88] We dream about things that have not yet happened.

But such links to the future are much rarer than links to the past. In dreams we are continually connected to our pasts, and dreams also take us into wider realms of possibility than our waking lives. Many things are possible in dreams that are not possible here.

Many people experience themselves as flying in their dreams. Some children dream that they can float up and down stairs, as I did, and some fly way beyond their own homes, as I did. Flying dreams seem to become less frequent as we grow up, judging by my own experience, and what my family members and friends have told me. You can find out for yourself. Think of your own experience. Did you have flying dreams as a child? Do you have them now? Ask people you know if they have flying dreams, or if they had flying dreams as a child. I suspect you will find that such dreams are fairly common, and more common in children than in adults.

Perhaps the experiences of flying children underlie baroque images of childlike cherubs with wings – flying children, including dead children, who flew away when infant mortality was high. There are flying children in classic fairy stories such as *Peter Pan*. In *Harry Potter* there are flying games on broomsticks. And what was once a fantasy is now mundane: anyone can fly in everyday life, thanks to planes.

Some people find that they can wake up within their dreams while still dreaming. They are dreaming and aware that they are dreaming. These so-called lucid dreams enable them to exert much more control over the dream than in an ordinary dream, sometimes including the ability to fly at will to places they want to visit. It is possible to cultivate the ability, and in the Hindu and Tibetan Buddhist traditions this is an ancient spiritual practice, called dream yoga.[89]

Many things happen in our dreams. Some dreams are scary, some comforting, some long, drawn-out anxiety scenarios, some blissful, some erotic, some visionary. And there are many kinds of beings in our dreams, including animals, other people, insects, monsters, dragons, deceased relatives, saints, angels, gods and goddesses.

Many of these dream themes are not based on our own personal experience. A study of the dreams of young children in New York City found that they had frequent nightmares that involved being chased by ferocious animals or monsters, rather than dreaming of the danger of being run over on a road, or other realistic hazards of modern urban life.[90] But their inherited dreams would have been very relevant to the experiences of children 100,000 or 1,000,000 years ago, when they lived in fear of tigers, cave bears or other predators.

The God of the Abrahamic religions sent messages through dreams and visions. He did so to Abraham (Genesis 15: 1). Abraham's grandson Jacob, also known as Israel, had a famous visionary dream in which he saw a ladder stretching between heaven and earth with angels ascending and descending on it, Jacob's ladder (Genesis 28: 10–19).

Joseph, the son of Jacob, rose to high office in the Pharoah's Egypt through his gifts in interpreting dreams, and was able to provide food for his eleven brothers when they and their father were starving in a famine. Skilful dream interpretation saved the lives of all twelve of Israel's sons, the patriarchs of the twelve tribes of Israel. Many years later, during the captivity of the Jewish people in Babylon, the prophet Daniel interpreted the dream of the King Nebuchadnezzar, and as a result rose to great power.[91]

In India, one of the many cosmological myths is that the god Vishnu lay down to sleep and dreamed. Our world is his dream. By dreaming his dream, Vishnu sustains the universe. When he wakes from his dream, a cycle of creation ends. The universe disappears. But he can dream another one.

In Europe 25,000 years ago, the imagery in the paintings on cave walls of animals and strange human-animal hybrids look like visionary creatures seen in psychedelic visions or in trances induced by drumming, chanting and dancing. At the very dawn of pictorial art, visions were turned into images. In turn, these images would have helped to fuel further visions and dreams in a shared mental realm.

Dreams take us into self-luminous mental realms that are more or less far removed from our waking lives and experiences. Psychedelics help propel our minds into highly visual realms that are akin to dreams and visions. In this context, it is not surprising that psychedelic experiences quite often involve encounters with seemingly autonomous beings, as well as travelling into other realms.

These realms are all ultimately made of mind, but not purely made of our own minds, or even of collective human minds, but of many kinds of minds. And, in the view of most religions, all these beings have a common source, a consciousness that lies behind them all, and sustains them all, the mind of God, or Brahman, or the All.

Morphic resonance and collective memories of psychedelic experiences

In my hypothesis of morphic resonance, I suggest that memory is inherent in nature and that all species have a kind of collective memory. If rats learn a new trick in London, rats of the same kind all over the world should be able to learn it quicker thereafter.

There is already evidence from laboratory experiments that this occurs.[92]

Morphic resonance depends on similarity. The more similar a self-organising system in the present to systems in the past, the stronger will be the resonance. From this point of view, when someone takes a psychedelic substance, say LSD, he or she will come into resonance with all those who have taken LSD before, because their brains, nerve cells and neurotransmitter systems are affected in a similar way by the same drug. A present taker of LSD will potentially tune in to a collective memory of countless past takers of LSD and in turn contribute to this collective memory.

The same goes for magic mushrooms, or DMT, or peyote, or iboga. If I take any of these substances, my brain will be affected in a characteristic way, and my experience will involve tuning in to people who have taken that substance before. Each of these drugs will have its own kind of collective memory, a pool of experience from people who have taken it before. When drugs have been taken in shamanic cultures, then the experience will be heavily charged with archetypal patterns from that culture.

This hypothesis may help to explain the observations of the psychologist Claudio Naranjo, who gave ayahuasca to people living in urban settings in South America. Although these were middle-class people who knew nothing about tribal cultures, they had visions of jaguars and snakes, which are major features in the mythologies of the cultures that used this visionary brew traditionally.[93] It seemed as if they were tapping into a kind of drug-specific memory.

Similar effects occurred with magic mushrooms in the 1950s. Albert Hofmann was asked to identify the active principles in a sample of psychoactive mushrooms from Mexico, and took some himself. 'Whether my eyes were closed or open, I saw only Mexican motifs and colours. When the doctor supervising the experiment bent over me to check my blood pressure, he was transformed into an Aztec priest and I would not have been astonished if he had drawn an obsidian knife.'

Hofmann knew the mushrooms were from Mexico, which could have influenced his experiences, but he found that others who ate the mushrooms without knowing where they came from would often see ancient Mexican art images.[94]

In principle, the morphic resonance hypothesis is experimentally testable. As Sasha Shulgin pointed out, when he made a new drug and took it for the first time, a new experience came into being that no human had ever had before. When his tasting panel took it, a consensus began to emerge about its effects. I suggest that this consensus reality could be influenced experimentally, to find out whether this would affect people who took it subsequently.

For example, there could be six possible settings in which the first group of people could take the drug: an English-style garden, a Japanese-themed room, a room with African décor and music, and so on. Then one of these six possible settings would be chosen at random, by the throw of a die. If, for instance, the die selected the Japanese setting, then the first twenty or thirty people would take the drug in that setting. Subsequently, different people who knew nothing of this previous history would take the drug in a neutral, featureless setting. If they started experiencing Japanese-style imagery, this would support the morphic resonance hypothesis.

Religious arguments against psychedelics

Some followers of traditional religions are opposed to the use of psychedelics for at least four reasons:

1. These substances are illegal. But the situation is changing. Now that the ritual use of ayahuasca and peyote is legal in some

parts of the world, the illegality argument is only applicable in some places, not everywhere.

2. Psychedelics can be harmful to some people. For children and for some people with mental-health problems, psychedelic experiences are indeed inadvisable. But their effects in a ritual context and when used over long periods of time are matters for research, rather than speculation. There have already been several studies of members of psychedelic churches. One such study in urban and rural settings in Brazil concluded, 'Overall, the ritual use of ayahuasca . . . does not appear to be associated with the deleterious psychosocial effects typically caused by other drugs of abuse.'[95]

A study in Spain compared regular users of ayahuasca in the context of psychedelic churches with regular practitioners of other religions. The ayahuasca users had lower scores on all psychopathology measures, and higher scores on the Spiritual Orientation Inventory, the Purpose in Life Test and the Psychosocial Well-Being Test. In a follow-up study a year later, the researchers wrote, 'We found no evidence of psychological maladjustment, mental health deterioration or cognitive impairment in the ayahuasca-using group.'[96]

3. Using drugs is spiritual cheating, trying to take a shortcut to visionary and mystical experiences.[97] But other practices can also lead to visionary experiences, including fasting, sleep deprivation and long periods spent in darkness. Are these forms of cheating, too? Or are they more valuable because they take more effort? Then what about spontaneous mystical experiences, which seem to take no effort at all, but happen as a gift or a blessing? It is hard to put forward an argument that validates some kinds of mystical experience, but invalidates others on the basis of effort.

4. Taking psychedelics is an entertainment, like going to the cinema, or enhancing celebrations. These drugs can make people feel happy. They can also make them feel they are having cosmic insights. But what good do these experiences do? How do they affect the life of the person who experiences them, and the lives of those around them? Unless these experiences change people's lives for the better, they are not helping their spiritual development. They are not spiritual experiences, but just trips. This argument may be true for some people. But for others, psychedelic experiences seem to provide spiritual openings, and change their lives for the better.

Two practices with mind-altering substances

I cannot advocate breaking laws that prohibit the possession and use of cannabis and psychedelics. However, I suggest the following practices for people who live in places where the use of these drugs is legal.

CANNABIS IN HOLY PLACES

There are many kinds of holy place, including sacred groves, healing springs, sacred mountains, temples, shrines, cathedrals and churches. Many people have found that visiting such places after taking cannabis has led to an enhanced sense of connection with the spiritual power of the place, and a greater ability to be present within it.

I do not recommend this practice for people inexperienced with cannabis, in case they take an inappropriately high dose or become fearful or paranoid.

To make this a spiritual practice and not just a drug experience, I suggest the following steps:

- Treat your journey to the holy place as a pilgrimage (*SSP*, chapter 7). Arrive on foot, even if you only walk the last half-mile. Go with the intention of giving thanks, and asking for healing, blessing or inspiration.
- When you take the cannabis, before arriving, give thanks, and ask for your visit to be blessed.
- If possible, walk around the place before entering it. In India, it is a standard practice to circumambulate a temple before entering it, walking around it clockwise, in the direction of the sun through the sky. Walking around a holy place has the effect of making that place the centre. In some cultures, like the Bon in Tibet, and Muslims in Mecca, circumambulation goes anticlockwise, so there is no absolute rule. I myself follow the Indian system and walk clockwise.
- When you enter the place, give thanks, and open yourself to its power and its blessings.
- Before you leave, give thanks for your experiences there, make an offering or donation, and ask for the blessings you have experienced to help you in your own life, and help you to help others.

PRAYER AND PSYCHEDELICS

If you want to go on a psychedelic journey, treat it as a kind of pilgrimage.

Make sure that you do so in a safe place, with supportive people around. Do not do it in a chaotic environment or a place you feel unsafe.

You are going where you are not always in control. This can be frightening. If you believe in a power greater than yourself, then ask for this power's protection, help and guidance. If you believe in God, ask for God's protection, help and guidance. If you only believe in yourself and your own power, then ask your higher self to watch over you.

Give thanks for your chance to do this, and ask for protection and blessings.

If your experience is challenging, again pray for help and protection.

And after your journey, give thanks for what you have seen and learned, and ask that it will help you and help others.

5
Powers of Prayer

Almost everyone can meditate. Not everyone can pray.

Traditionally, people have meditated so that they can come into connection with a consciousness far greater than their own (*Science and Spiritual Practices*, chapter 1). But in our modern secular world, many materialists and atheists meditate without believing in any greater consciousness. They believe that the effects of meditation are all inside their bodies, and especially inside their heads.

By contrast, prayer presupposes a relationship with more-than-human consciousness. It would make no sense if it were confined to the heads of the people praying. Prayer is explicitly addressed to gods, goddesses, spirits, ancestors, saints, angels and God. For an atheist and materialist, all of these entities are non-existent; they are projections of human minds. They are not really 'out there'. They are inside heads. There is no point in praying to an imaginary being.

In order to pray, it is necessary to believe that there are beings to whom prayer can be addressed, and that these beings can help. In all traditional societies, this belief is taken for granted. But in more complex societies, it became possible to doubt or deny the existence of such beings. For example, within the Roman Empire many educated people were sceptical about the official gods. In this context, trust in the existence of spiritual beings and in their power became a prerequisite for prayer.

In the Bible, this principle is explicit. You cannot pray to God if you do not believe he exists and responds to prayer: 'Whoever would approach him must believe that he exists and that he rewards those who seek him.' (Hebrews 11: 6).[1]

Thus the practice of prayer is a kind of litmus paper for world-views. For most atheists and materialists, prayer is pointless, and a very small minority of non-religious people pray regularly.[2] But for most people, prayer is meaningful; there are conscious beings who can be prayed to. In shamanic cultures, these beings include animal spirits, ancestors, Mother Earth and a Great Spirit who includes all things. For Hindus, prayers can also be addressed to a wide range of gods and goddesses.

Buddhists do not like to bother the serenity of Buddha with mundane requests, so in Buddhist countries, prayers about practical affairs, healing, protection from enemies, success in business or exams, or having children, are addressed to a range of gods, goddesses and guardian spirits, who are often versions of older, pre-Buddhist deities. Orthodox and Catholic Christians can pray to saints and angels, asking for their intercessions. And Jews, Christians and Muslims can all address their prayers direct to God, as do many Hindus, who see an ultimate reality beyond all gods and goddesses, often described as *sat-chit-ananda*, being-consciousness-bliss.

Kinds of prayer

There are many kinds of prayer. Some are prayers of thanksgiving and gratitude (*SSP*, chapter 2). Contemplative prayer in the Christian mystical tradition is very similar to what is now usually called meditation. In many other traditions, some form of prayers, like meditation, involve the repetitive mantra-like words or phrases, as in the chanting of *Hare Krishna* by Hindu devotees, or the repeated saying of the Jesus Prayer by Eastern Orthodox Christians. Some prayers involve the repetition of prayers or phrases using prayer beads or *malas*, which are used by millions of Buddhists, Hindus and Sikhs; Muslims use prayer beads called *tasbih* or *misbaha*; Eastern Orthodox Christians use prayer ropes; and Roman Catholics use rosaries.

Many prayers are prayers *for* something. They include requests for healing, or protection, or blessings, or good fortune, or success. Such prayers are called petitionary prayers. Some are prayers in which people pray for themselves. When they pray for others, they are making intercessory prayers. And many people pray on behalf not only of other people, but also for institutions like schools, hospitals, churches, teams, societies and governments; or for animals, or gardens, orchards and fields, or for local community, for the nation, or for the world. All these are kinds of asking prayers, petitionary prayers.

In this chapter, I focus on petitionary prayers, prayers that are asking for something. After a brief discussion of the worldwide prevalence of prayer, I look at the assumption on which prayers are based – namely that our minds are transparent or porous, and that beings greater and more powerful than ourselves can hear our prayers, and be influenced by them. I then discuss prayers for healing, and scientific studies of the effects of prayer on health and wellbeing. But prayers also have a shadow side; the power to bless is also the power to curse. I then discuss the difference between prayer and positive thinking.

I myself both pray and meditate. I meditate in the mornings soon after I get up. I pray in the evenings just before I go to bed. I think of meditation as like breathing in, taking the focus of concentration inwards; prayer is like breathing out, focusing concentration on blessings, purposes, goals, intentions and hopes. Meditation is not about requests or intentions, but rather about centring on the process of breathing and bodily sensations, as in mindfulness meditation, or on the repetition of a mantra.

During meditation, thoughts and sensations come and go, and by returning to the awareness of breathing or the mantra, there can be moments in which the meditator feels part of a much greater presence, no longer absorbed in internal dialogues about the past and the future. The aim of meditation is to let go of thoughts without engaging with them.

Petitionary prayer involves a flow in the opposite direction. It starts by invoking the spiritual being to whom prayers are addressed, and then focuses attention on specific people, needs or requests, directing intentions outwards.

The worldwide practice of prayer

Over a wide range of societies, prayers are the expressions of desires cast in the form of requests to spirit beings – beings with the power to answer. In many cultures, prayers are made only intermittently, as and when needed to deal with fertility, childbirth, illness, earthquakes, droughts, and dangers. Prayers are often thought to be especially important on the occasions of death and funeral ceremonies, when potentially dangerous influences are more intimately in touch with human life.

In some cultures, one of the functions of funerals is to ensure that the dead pass on to the appropriate realm of the dead, rather than remaining in the world of the living as ghosts.[3] In weddings, there are often prayers for children. In agricultural societies, prayers are often offered at the times of sowing and harvesting of crops.[4] These kinds of prayers are primarily concerned with survival and material needs, rather than spiritual benefits. They are not focused on ethics.

In many societies that recognise the existence of a supreme spirit, offerings and prayers are usually made to lower-level spirits, who are thought to be more concerned with human affairs. For example, ancestral spirits or saints are asked to intercede with the supreme spirit, or animals are sacrificed and charged with carrying a message to the gods.

Among the ancient Greeks, as in many other cultures, different gods or goddesses had different areas of responsibility: poets prayed to the muses, hunters to the hunting goddess Artemis, and farmers to the corn goddess Demeter. Most Greek prayers were essentially utilitarian, for health and wealth, children, and success in business and in battle, and the same is true in many other parts of the world.

Often prayers are made by specialists, like shamans, who know how to invoke the spirits, or by priests, or by leaders who act as representatives of the community as a whole, or by heads of households.

In some cultures, prayers are part of regular life, every morning or evening, or at new moon, or at other regular times.

In traditional societies, religion is primarily practical rather than theoretical; it is not about a well-defined belief system, but about fulfilling needs. The gods, spirits and ancestors are interactive, and people interact with them by giving, receiving, promising, threatening, and placating.[5]

One aspect of their interactive nature, taken for granted by most people, is their ability to understand human languages. In many parts of the world, people traditionally pray out loud; but when people pray silently, they implicitly assume that the being to whom they are praying not only understands their language, but can hear it when it is expressed silently. They assume that their minds are transparent to gods, spirits and ancestors.

The practice of prayer is far more widespread and far more ancient than the development of theology. As one scholar of comparative religion put it, 'Prayer may truly be said to be prior to all definite creeds, to be indeed the expression of the need which all creeds seek to satisfy.'[6]

In the monotheistic religions, many of these general principles about prayer apply to prayers addressed to God. But there are important differences between praying to ancestors, spirits, saints, angels, gods and goddesses on the one hand, and, on the other, God. First, belief in a supreme deity creates a greater sense of security than the uncertainties and confusions of a polytheistic pantheon. God's power has no rival among other gods or demons.

And secondly, starting with the Jewish people, a belief in God's ultimate holiness means that only through moral goodness can people become acceptable in his sight.

Jesus taught a new and more intimate way of praying to God, addressing him less as an all-powerful ruler and judge, and more as a loving father. In prayer, God was to be approached as if by a child, in simplicity and directness, in confidence and love. Prayer should not be done for outward show, or as a magical incantation. And as well as prayers in public, Jesus advised his followers to pray privately: 'Go into your room and shut the door.' (Matthew 6: 7).

The Lord's Prayer, given by Jesus, sets a pattern for Christians. Prayers for the coming of God's kingdom precede the prayers for daily needs and for forgiveness. The supreme end, the universal good, comes first:

Our Father, who art in heaven, hallowed be thy name.
Thy kingdom come; thy will be done on earth as it is in heaven.
Give us this day our daily bread,
And forgive us our trespasses, as we forgive those who trespass
  against us.
Lead us not into temptation, but deliver us from evil.
For thine is the kingdom, the power and the glory, for ever and
  ever. Amen.

In Muslim prayer, too, the supremacy of God comes first, and the prayer depends on following the right path in life. It is not for selfish ends. One of the regular daily prayers is as follows:

Praise belongs to God, Lord of the Worlds,
The Most Kind and Merciful,
Master of the Day of Judgement.
It is you we worship; it is you we ask for help.

Guide us on the straight path: the path of those you have blessed,
Those who incur no anger and who have not gone astray.
Ameen.[7]

Among Christians, although all prayer is ultimately addressed to God, there is a continual tendency for people to pray to saints as intercessors, feeling that they are closer to human concerns and also closer to God than the person praying. They are intermediaries. Prayer to saints is a major feature of the Eastern Orthodox and Roman Catholic traditions, and churches and cathedrals contain many icons and images of these saints, as well as chapels dedicated to them. The saints have now migrated online. Details of hundreds of saints, including information about their feast days, are to be found on Catholic websites, where it is also possible to light slow-burning, virtual prayer candles to them. Each candle lasts a year, and costs $15.[8]

In addition, prayers are often addressed to angels. A Roman Catholic prayer to one's guardian angel runs as follows:

Angel of God, my Guardian dear,
To whom his love commits me here,
Ever this night be at my side,
To light and guard, to rule and guide.[9]

At the Reformation, Protestants rejected prayers to saints and angels, and suppressed pilgrimages to their shrines. Likewise, within Islam, where prayers to Sufi saints are widespread, there was and still is strong opposition to the cults of saints and pilgrimages to their shrines, most notably by followers of the Wahhabi movement, the dominant form of Islam in Saudi Arabia.

Within the Judeo-Christian tradition, there has always been a tension between the direct path to God at the core of the prophetic tradition and the attractions of sacred groves, seasonal festivals, ancient stones, goddesses, the shrines of saints and local ancestor cults. The Jewish prophets continually inveighed against them. But they co-existed in the Holy Land.[10]

Within Christendom, the Eastern Orthodox and Roman Catholic traditions established long-term compromises. Many aspects of old faiths, festivals, rituals and traditions persisted within a Christian framework, and at the same time, the direct path to God was kept open by professional contemplatives, especially monks, nuns and hermits.

At the Reformation, in the sixteenth century, Protestants like Martin Luther and John Calvin denounced this compromise. The direct path to God was not for the few, a monastic spiritual elite, but for everyone.

This was a high ideal. Not everyone feels a calling to this dedicated spiritual path. And when Protestants are not drawn to it, or turn away from it, then they do not have the saintly and angelic helpers of the Orthodox and Roman Catholic churches to fall back upon. They cannot seek the help of intermediaries who seem more approachable. It's God or nothing.

In modern evangelical churches, God is experienced through prayer with an extraordinary intensity and immediacy. Pastors and leaders advise people to pay close attention to the thoughts and images that come into their minds, so that they can learn to recognise God's voice in everyday thoughts, and also to play close attention to their dreams and visions. They also attend to patterns and events in their lives that sceptics would dismiss as more co-incidences.[11]

The anthropologist Tanya Luhrmann made a pioneering study of modern American evangelical Christianity that she summarised in her book, *When God Talks Back*. For these Christians:

Religion is not about explaining reality, but about transforming it: making it possible to trust that the world is good, despite ample evidence to the contrary, and to hope, despite loneliness and despair . . . What they want from faith is to feel better than they did without faith. They want a sense of purpose; they want to know that what they do is not meaningless; they want trust and love and resilience when things go badly . . . The evangelical Christianity that emerged out of the 1960s is fundamentally therapeutic. God is about relationship, not explanation, and the goal of the relationship is to convince congregants that their lives have a purpose and that they are loved.[12]

Just as prayer can reinforce the experience of a connection to more-than-human consciousness, not praying can reinforce the experience of a disconnection from more-than-human consciousness, and strengthen the belief that we live in a world of purposeless matter. In the modern secular world, the default assumption of most non-religious people is that the universe is non-conscious and that there is no such thing as divine intervention. Humans are able to control the environment with no need for help from anything outside themselves, and with no possibility of such help. We humans are on our own. The universe has no ultimate purpose or meaning.[13]

One of the practical consequences of this atheist or materialist or secular humanist worldview is a disbelief in prayer. Most atheists do not pray, and feel they cannot pray. Either prayer would be pointless or it would be hypocritical.

In a survey in 2016 of scientific, engineering and medical professionals in Britain, about forty-five per cent of the respondents described themselves as non-religious, a term that included both atheists and agnostics. An equal proportion, forty-five per cent, described themselves as religious or spiritual. Both religious/ spiritual and non-religious people meditated. Among the religious/

spiritual, twenty-two per cent meditated at least once a month, as did sixteen per cent of the non-religious. In relation to prayer, the difference was much greater. Among the religious/spiritual, fifty-one per cent prayed at least once a month, compared with less than one per cent of the non-religious.[14]

Among the very small minority of atheists that pray in spite of their beliefs is the neuroscientist Todd Murphy. Like most other atheists, he believes that all his experiences are confined to the inside of his head. There is no God 'out there'. He thinks that the sense of presence he experiences in prayer is a part of his own mind, produced by the right hemisphere of his brain:

I am an atheist who prays. By extension, I'm an atheist with faith in God (or the Gods). Intellectually, I understand that the God of my prayers is an expression of my own right hemisphere sense of self. As a point of faith, I understand that this, my own personal God, is a presence I *sense* . . . My prayer life is fraught with contradictions, but I believe I am explicit enough here not to be considered a hypocrite. My beliefs and feelings contradict each other, and I think that's the way it's supposed to be, because humans did not evolve for truth, but rather for survival, and letting go of stress feels good.[15]

Murphy is very unusual in living with such extreme contradictions between materialist beliefs and his prayer life. Most atheists and agnostics do not pray. Most religious or spiritual people do.[16]

Transparent minds

In traditional societies all over the world, people take it for granted that their minds are transparent to gods and spirits. As the anthropologist Pascal Boyer put it:

God knows more than we know, the ancestors are watching us. More generally: In most local descriptions of spirits and other such agents, we find the assumption that they have access to information that is not available to ordinary folk. What is made explicit is most often a vague assumption that the spirits or the gods simply know more than we do. But it seems that people in fact assume something more specific, namely that the gods and spirits have access to *strategic* information rather than information in general.[17]

Strategic information is information relevant to social interactions, like knowing people's hidden intentions, and whether they are lying or cheating. Most people are not concerned with the more abstract question of whether gods or spirits know *everything*; they simply assume that these beings know what is relevant to the situation being prayed about. For example, spirits may or may not know about every item that is in your house, but if you stole one of these items, then they will be aware of that fact.

The assumption that our minds are transparent also underlies the belief that God, saints and spirits can hear our prayers wherever we are, regardless of whether we are with other people or on our own, sitting on a train, driving a car, in a hospital bed, or in a remote forest. They can hear silent prayers inside our minds, not only prayers spoken out loud, and they can understand all human languages. We can hide nothing.

These assumptions are implicit in all traditions, and are sometimes fully explicit. Among the Kwaio people of the Solomon Islands, the ancestors, the *adalo*, 'see the smallest slight things. Nothing is hidden from the *adalo*.'[18] One of the fullest and most poetic statements of this belief in the Jewish scriptures is in Psalm 139:

O Lord, you have searched me and known me.
You know when I sit down and when I rise up;
You discern my thoughts from far away . . .
Even before a word is on my tongue,
O Lord, you know it completely . . .
Where can I go from your spirit?
Or where can I flee from your presence?
If I ascend to heaven, you are there;
If I make my bed in Sheol [Hell], you are there.
If I take the wings of the morning
And settle at the farthest limits of the sea,
Even then your hand shall lead me . . .
For it was you who formed my inward parts;
You knit me together in my mother's womb . . .
My frame was not hidden from you,
When I was being made in secret,
Intricately woven in the depths of the earth.[19]

Likewise, one of the prayers in the Anglican *Book of Common Prayer* begins, 'Almighty God, unto whom all hearts are open, all desires known and from whom no secrets are hid . . .' In the Koran, 'Whether you conceal what is in your hearts or bring it into the open, God knows it.' (3: 29).

People who pray retain a sense that their minds are porous; they are open to spiritual influences that come from outside, both good and bad. They are not only open to the spirits of God, the saints and angels, but also potentially open to the harmful influences of demons and destructive spirits. In Europe up until around the year AD 1500, almost everyone experienced themselves as porous in these ways. But through a long process of secularisation, a new sense of selfhood developed. Individual minds were progressively separated from each other and from spiritual influences. The philosopher Charles Taylor called this new sense of self the 'buffered self', enclosed in barriers that separate off individual minds.[20]

This separated self reaches its fullest development in the modern materialist view of the mind as nothing but an activity of the brain, confined to the inside of the head, insulated within the privacy of the skull. This is the default assumption of most atheists. They believe that their thoughts and desires are private, and that their minds are separate and autonomous. This is one reason why many militant materialists are so opposed to the existence of psychic phenomena, and why they deny the very possibility of telepathy. If they admitted that other people's thoughts and intentions could directly affect their minds, or that their own thoughts and intentions could be detected by other people, then they would be admitting to a porousness that would undermine their isolated selves.

This separated selfhood takes time to develop in modern people as they grow up. A baby has no secrets, and there is a sense in which it is completely open to its parents. It does not have the idea of a separated ego that is private. And many children take it for granted that God can see everything – including things that their parents cannot see.[21]

With the rise of secularism from the eighteenth century onwards, more and more people experienced their minds as enclosed within their heads, no longer transparent to God or other spirits. But at the same time, they imagined their minds being transparent to other humans in a new way, through novels. The novelist became like a god or a spirit who could see into people's minds. We now take it for granted that a novelist can read minds and tell us what people in the novel are thinking, feeling, desiring and fearing. Within the pages of novels, the old idea of transparent minds takes on a new lease of life.

Finally, through modern technology, our private secular selves are once again porous and open. Almost no secrets are hidden.

Our Google searches, our emails, our social-media communications, and our telephone calls are transparent to corporate and state surveillance; our smartphones transmit our exact location. Our Fitbits record our sitting down and our rising up; they monitor when we sleep deeply and when we dream; and they place all this information in cloud storage systems.

Scans of the internal workings of our brains and bodies are kept in our searchable medical records. Scans of unborn babies in their mothers' wombs show the details of the developing bodies. Embryos are turned into computerised images. Not only are many babies scanned before they are born, they now grow up amidst these omniscient technologies. They may have less sense of a private self than their parents.

A secularised version of Psalm 139 is coming true. But unlike God, these secular agencies are not all benevolent. Some are indeed concerned with our health, wellbeing and security, but most are about monetising information.

Praying for healing

Many people pray for healing, for their families, friends, animals and themselves. Obviously, not all prayers for healing work. Not everyone prayed for gets better. Everyone ultimately dies, so prayers for limitless healing are doomed to failure. For the same reason, all medicine is eventually doomed to failure. It cannot heal forever, or conquer death. Prayer and medicine work sometimes, but not always.

Praying for healing probably occurs in all cultures and religions. Even in the context of modern medicine, such prayers are very common. A survey in 1998 found that seventy-five per cent of Americans had prayed for their own health. A survey in 2002 by the US Centers for Disease Control and Prevention found that prayer was the most common form of alternative therapy, and that during the previous year, forty-three per cent of Americans had prayed for their own health and twenty-four per cent had asked other people to pray for them.[22]

Some churches place a strong emphasis on prayers for healing, especially Pentecostal Christians, who have a particularly strong faith in spiritual gifts, and especially divine healing. From a small handful of adherents at the beginning of the twentieth century, Pentecostalism has grown through a series of revival movements and is now extended all over the world, most dramatically in Latin America, Asia and Africa.

Studies of the global Pentecostalist movement by the ethnographer Candy Gunther Brown and her colleagues have shown how this movement spreads. In her book, *Testing Prayer: Science and Healing*, Brown summarised this process as follows:

> Those who experience healing attribute their recoveries to divine love and power and consequently feel motivated to express love for God and other people in part by praying for others' healing . . . Such efforts are self-perpetuating, because individuals both disburse and draw emotional energy – or love energy – from social interactions in which recipients, partners and leaders all experience love through their involvement in healing rituals. Because Pentecostals perceive God to be an unlimited source of love and power, they are emboldened to expend rather than conserve energy.[23]

These prayers are sometimes followed by instant cures, which are regarded as miraculous. But often the distinctions between the effects of prayer and conventional medicine are blurred. Pentecostalists and many other Christians pray for God to guide the hands of surgeons, for medicines to work more effectively, or for the Spirit to work through psychological counselling.[24]

In the early twentieth century, sceptics who wanted to discredit claims of religious healing proposed that medical records should be used as the standard for determining whether healings had in fact occurred. From the 1960s onwards, American Pentecostalists took up this challenge, and collected before-and-after medical documentation of attested healings. The data running from the 1960s to 2011 showed that 'some, though not all, individuals attesting to religious healing exhibited medically surprising recoveries. Medical records point to resolution of a wide variety of diseases and disabilities, including severe auditory and visual impairments and metastasised cancers.'[25]

There is also a strong tradition of healing through prayer among Roman Catholics. Many saints are revered for their healing powers, and one of the criteria for the canonisation of saints is the performance of miracles, usually healing after their death. Many people go on pilgrimages to healing shrines, which are often springs or holy wells. The most famous is at Lourdes, in the foothills of the French Pyrenees. Following a series of visions of the Blessed Virgin Mary by a fourteen-year-old peasant girl, Bernadette, in 1858, miraculous healings began when people drank the water from a spring that Bernadette unearthed on the instructions of the Blessed Virgin.

Today thousands of gallons gush out from this source, and pilgrims bathe in the waters. Lourdes now attracts more than six million pilgrims and visitors a year. Countless people claim to have been cured there, and the Lourdes Medical Bureau provides detailed documentation of exceptional healings that are officially classified as miracles.[26]

Another way of investigating the healing power of prayer is through prospective studies, in which people are studied before the healing procedure and examined again afterwards. In one such study, Brown and her colleagues worked with people with seeing and hearing problems in rural Mozambique. They measured their visual and auditory thresholds before and after they were prayed over for healing, and found 'highly significant improvements in hearing and statistically significant improvements in vision across the tested populations.'[27]

In all these cases, people knew that they were being prayed for, or that they were at a healing shrine. But what happens if people do not know they are being prayed for, and if the people praying for them are strangers taking part in a scientific experiment?

In several scientific studies of the effects of healing at a distance, the people praying were nowhere near the people being prayed for, sometimes hundreds of miles away, and they did not know them. These studies were carried out using standard double-blind methods, as in clinical trials of new drugs. One randomly selected group of people was prayed for and another group was not prayed for. The subjects did not know whether they were being prayed for or not, and neither did the people looking after them. The people who did the praying were strangers.

Some of these studies gave positive, statistically significant results, but the largest and best organised, coordinated by researchers at Harvard Medical School, showed no significant beneficial effects. This experiment included 1,802 patients undergoing coronary bypass surgery in six hospitals in the US. One Protestant and two Catholic groups prayed for selected patients to have a quick recovery with no complications. They were provided only with the first name and initial of the last name of the patients, such as 'Mary B'.

The patients were divided at random into three groups. One group was told they might or might not be prayed for and was prayed for. Another group was told the same, and was not prayed for. The third group was told that they would be prayed for, and was prayed for. When the results were analysed, there was no difference between the first two groups, but the third group had significantly *more* complications.[28] Praying seemed to have made them worse. The researchers themselves offered no explanation, other than that it 'might have been a chance finding.'[29]

I discuss these puzzling results further in the text box below.

***

The results of the Harvard prayer study show how complicated research on prayer can be, especially by applying standard drug-testing type protocols that seemed very unnatural to the participants in the study. The Harvard study showed that prayer had no effect in the groups that did not know if they were being prayed for, whether they were prayed for or not.

This is what sceptics would have predicted: there would be no effect of prayer. But then why did prayer have a significant *negative* effect on people who knew they were being prayed for, which no one had predicted?

If prayer in itself had negative effects, then the prayed-for double-blind group should have done worse than the control group. But there was no difference between them. So, the difference seems to be related to the fact that one group knew they were being prayed for and the others did not.

Perhaps the people that knew they were being prayed for thought that they were in exceptional danger. As a cardiologist on the research team put it, 'It might have made them uncertain, wondering, "Am I so sick that they had to call in their prayer team?" We found increased amounts of adrenaline, a sign of stress, in the blood of patients who knew strangers were praying for them. It's possible that we inadvertently raised the stress level of these people.'[30]

Why was there no difference between the groups that were prayed or not prayed for, without knowing which? At first sight, the results seem to show that prayer does not work. But as Stephan Schwartz and Larry Dossey have pointed out, both the prayed-for and the not-prayed-for group were probably being prayed for.

Many Americans pray regularly when they are well, according to surveys, 'and it can be assumed that even more pray when they are sick. Faced with the prospect of being denied prayer in this study, the subjects . . . may therefore have aggressively solicited prayer from their loved ones to make up for the possible withholding of prayer in the experiment, and they may have redoubled their personal prayers for themselves.'[31]

In retrospect, the design of the Harvard study was naïve. It assumed that the subjects in the experiment that they selected to be not prayed for were not being prayed for at all. They assumed that they would not be prayed for by their families, friends and churches, and that they would not be praying for themselves. But this was a flawed assumption. Most people in these American studies would probably have prayed for themselves and been prayed for by others.

The experimental design assumed that if prayer works, then the addition or withholding of prayer from strangers far away – who did not even know the full names of the people they were praying for – would have had significantly more effect than the prayers of people's nearest and dearest, their religious communities, and their own prayers. This is implausible.

What about the placebo effect? When people know that their family and friends are praying for them, and when they are praying for themselves, do these prayers serve as placebos?

The placebo effect occurs when people get better after they feel they are being cared for and given a healing treatment, even if the treatment is a blank pill. If prayer acts as a placebo, then atheists and sceptics can have no problem with its working. Placebo effects are well known in mainstream medicine, and are a major feature of clinical research using the so-called 'gold standard' of double-blind, placebo-controlled trials. They are 'double blind' because neither the patients nor those caring for them know who has received the treatment being tested, and who has received the placebo.

Patients who receive blank pills, thinking that they may be getting a new wonder drug, often get better. So do people who are given sham surgery instead of a proper operation. Their positive beliefs and expectations help to heal them. They are hopeful rather than despairing.[32] And prayers from loved ones are likely to make people feel more loved and more hopeful.

Jo Marchant, a former editor at the scientific journal *Nature*, made an extensive study of placebo and mind-body effects in healing, which she summarised in her book called *Cure: A Journey into the Science of Mind Over Body*. She describes herself as sceptically minded and non-religious. After studying the placebo response, examining many kinds of mind-body medicine, and working as a volunteer at Lourdes, she concluded:

> At the heart of all the pathways I've learned about is one guiding principle: if we feel safe, cared for and in control – in a critical moment during injury or disease, or generally throughout our lives – we do better. We feel less pain, less fatigue, less sickness. Our immune system works with us rather than against us. Our bodies ease off on emergency defences and can focus on repair and growth.[33]

In this context, it is not surprising that prayer often helps people get better.

If prayer helps people to feel safe, cared for and in control and thereby leads to placebo responses that facilitate healing, then prayer works, whether it is called a placebo response or an answer to prayer. Those who do not pray, and who are not prayed for, will not receive this boost to the healing process.

Sometimes the discussion about the healing effects of prayer is polarised into an either/or argument – *either* people should be prayed for *or* they should be treated medically, as if these are mutually exclusive. This may be the case for some religious groups that shun medicine entirely and rely only on prayer, or for atheists that reject prayer as unscientific, but for most people there is no either/or opposition: prayer and medicine can work together. For millions of people this is common sense.

In an extensive study in a National Health Service hospital in England, patients were offered healing sessions with trained spiritual healers in addition to their normal medical treatments. The patients were randomised into two groups: one group received healings while the other was put on a waiting list, and received healings after the main trial period was over.

The results were clear.[34] On average, patients who received spiritual healing sessions showed a significant reduction in their symptoms and improvements in their wellbeing, even though many of them were initially sceptical about the value of spiritual healing.[35]

Many doctors and nurses who pray feel inhibited by the secular culture of medicine about praying for their patients. The doctor and author Larry Dossey predicts that this situation will change. Instead of the default assumption that medical professionals should not pray for their patients, 'prayer will become recognised as a potent force in medicine and will become incorporated into the mainstream. The use of prayer will become the standard in scientific medicine in most medical communities.'[36]

So far this discussion has focused on physical healing, because this is the easiest kind of healing to study. Many people pray for emotional, mental and spiritual healing. Some people say they have received such healing. There seems no reason to doubt them.

Effects of prayer on health and happiness

In the late nineteenth and twentieth centuries, atheists attacked religion and religious practices on the grounds that they were irrational and harmful. Religion was 'neurotic', 'hysterical', 'delusional', had 'adverse effects on health' and even 'caused brain damage'.[37] Sigmund Freud compared religious people to the mentally retarded, and described religion as 'the universal obsessional neurosis of humanity.'[38]

Albert Ellis, an American psychologist, thought that 'sane and effective' psychotherapists should not go along with their patients' religious beliefs, because these beliefs are an 'emotional illness' that 'keep them dependent, anxious, and hostile, and thereby create and maintain their neuroses and psychoses.'[39] The Canadian psychiatrist Wendell Watters argued that Christian doctrines and teachings are incompatible with sound mental health, causing depression, reducing the ability to cope and even leading to physical health problems.[40]

These negative views were not based on scientific evidence and careful research, but on personal prejudices. But the question need not rest on personal opinions any longer, because there have now been several thousand scientific studies, published in peer-reviewed journals, on the effects of spiritual and religious practices. Over 2,800 such studies were summarised in the second edition of the monumental *Handbook of Religion and Health* (2012). A small minority (four per cent) reported that religious or spiritual people experienced poorer mental health, and eight and a half per cent reported poorer physical health; but the great majority showed a wide range of positive effects compared with matched groups of non-religious people.

In summary, these studies found that religious and spiritual (R/S) people experience more positive emotions (wellbeing, happiness, life satisfaction), fewer emotional disorders (depression, anxiety, suicide, substance abuse), more social connections (social support, marital stability, social capital), and live healthier lifestyles (more exercise, better diet, less risky sexual activity, less cigarette smoking, more disease screening, better compliance with treatment). This helps to explain why R/S people on average are physically healthier (less cardiovascular diseases, better immune and endocrine functions, perhaps less cancer, and better prognosis, and greater longevity overall).[41]

As well as studying the effects of religious practices in general, more specific studies have looked at the long-term effects of prayer on health and survival of people who pray. These studies were prospective; in other words, at the start of the study two groups of otherwise similar people were identified, one consisting of people who prayed regularly, and the other of people who did not pray. Both groups were watched over a period of years, keeping records of their health and mortality. Did they turn out differently?

They did. On average, people who prayed remained healthier and lived longer than those who did not.[42] For example, in a study in North Carolina, Harold Koenig and his colleagues tracked 1,793 subjects who were over sixty-five years old with no physical impairments at the beginning of the study. Six years later, those who prayed had survived sixty-six per cent more than those who did not pray, after correcting for age differences between the two groups. (Without this correction the difference was seventy-three per cent.)

Koenig and his colleagues then examined the effects of 'confounding variables', a scientific term for other factors that might have influenced survival, like stressful life events, depression, social connections and healthy lifestyles. Even after controlling for these variables, those who prayed survived fifty-five per cent more. They concluded that 'healthy subjects who prayed were . . . more likely to survive, and only a small percentage of this effect could be explained on the basis of mental, social, or behavioural factors.'[43]

If a new drug or surgical procedure had as dramatic effects on health and survival as praying, it would be hailed as a breakthrough – even as a miracle cure.

Prayer and positive thinking

Positive thinking is a kind of secularised prayer. By creating positive images of the future, visualising success, and cultivating an attitude of optimism, the positive thinker aspires to achieve success in love and business – or in more modest goals, like finding a parking place. But unlike prayer, positive thinking is not directed to a more-than-human consciousness, nor is it usually about helping others or doing God's will. It is about personal success and getting ahead.

The roots of the modern positive-thinking movement lie in the nineteenth century. One ingredient is the self-help tradition, made fully explicit by Samuel Smiles' book, *Self-Help*, first published in 1853, which spawned an enormous literature of self-help books, especially in the United States.

Another major ingredient is the New Thought Movement in the late nineteenth century.[44] This movement was intensely individualistic, as opposed to the community-based forms of traditional religions. In 1917, one of New Thought's proponents, Abel Leighton Allen, summarised it as follows: 'New Thought proclaims a robust individualism. The individual is the unit from which all greatness springs. Man is great only as he is individualistic, only as he follows his own path. The individual ranks above all institutions.'[45]

The God that New Thinkers pray to is not a God out there, but the God within themselves. They believe that mind is the dominant power in people's lives and that personalities are the sum total of their thoughts. Thus by positive thoughts they can move forwards and become part of the evolutionary process. As Allen put it, 'New Thought means spiritual and mental growth, constant and eternal progress. Recognising divine qualities in man, it sets no bounds to the soul's progress.'

One of the most influential books in this tradition was *The Power of Positive Thinking* (1952), a bestselling self-help book by Norman Vincent Peale. This book encourages the reader to achieve a permanent constructive and optimistic attitude by using affirmations and visualisations. Peale had an enormous influence within the United States. He was a personal friend of President Richard Nixon. Ronald Reagan awarded him the Presidential Medal of Freedom in 1984 for devising 'a philosophy of happiness'.[46] He was the pastor of a Reformed Church in New York from 1932 until his death in 1993. Donald Trump attended Peale's church while he was growing up, and Peale was the minister at Trump's first wedding, to Ivana.

Peale's book is replete with stories illustrating the successes of positive thinking, such as the story of a travelling salesman who made use of cards with inspirational texts. He told Peale, 'Since I've been using these cards as I drive and committing the words to memory, the old inferiority and disbelief have gone. It's really wonderful the way this method has changed me. It has helped my business, too.'[47] Peale advised his readers:

Trust yourself. You cannot be successful or happy without confidence in your own powers and abilities. To succeed you need self-confidence . . . Such a mental attitude is extremely important and this book will help you believe in yourself and release your inner powers . . . 'Attitudes are more important than facts.' That is worth repeating until its truth grips you. Any fact facing us, however difficult, even seemingly hopeless, is not so important as our attitude towards that fact.[48]

Sceptics denounced his evidence as merely anecdotal. Psychologists claimed that his techniques were based on a form of autosuggestion or self-hypnosis that could be harmful. And although Peale claimed these techniques would give those who practised them 'the power of God', many religious leaders felt that he was seriously misrepresenting Christianity. Nevertheless, many people welcomed his message and practised his methods.

One exceptionally important book in the New Thought Movement was *The Science of Getting Rich* by Wallace D. Wattles, first published in 1910, which was a major inspiration for a book and film called *The Secret* by Rhonda Byrne (2006), which re-popularised the New Thought principle of the 'Law of Attraction', namely the belief that focusing on positive or negative thoughts brings about positive or negative experiences through a principle of cause and effect.[49]

The school of positive psychology, which grew up within the mainstream academic psychology, overlaps with positive thinking in some ways, but proponents of positive psychology are keen to emphasise the differences. One of the leading lights of this movement, Martin Seligman, a former President of the American Psychological Association, contrasted positive thinking with the 'learned optimism' of positive psychology: 'Positive thinking often involves trying to believe upbeat statements such as "Every day in every way, I am getting better and better," in the absence of evidence or even in the case of contrary evidence . . . Learned optimism, in contrast, is about accuracy.'[50] Seligman argued that positive psychology is better than positive thinking because it is more scientific:

First, positive thinking is an armchair activity. Positive Psychology, on the other hand, is tied to a program of empirical and replicable scientific activity. Second, Positive Psychology does not hold a brief for positivity. There is a balance sheet, and in spite of the many advantages of positive thinking, there are times when negative thinking is to be preferred. Although there are many studies that correlate positivity with later health, longevity, sociability, and success, the balance of the evidence suggests that in some situations negative thinking leads to more accuracy . . . Positive Psychology aims for the optimal balance between positive and negative thinking.[51]

There are clearly similarities between the practice of prayer, positive thinking and positive psychology. They can all have positive effects on people who practise them. And sometimes the distinctions between positive thinking and prayer are blurred. Some Protestant churches in the US and elsewhere follow a doctrine of 'prosperity theology', which since the 1950s has had much influence on televangelism, and on the Pentecostal and Charismatic movements.[52]

The prosperity gospel is akin to New Thought in that it is based on the belief that financial blessings and physical wellbeing will be delivered to people who are personally empowered through faith, positive speech, positive thoughts – and donations to religious causes. This kind of theology has many critics in mainstream Christian churches, but is spreading worldwide, especially among poor people. It gives them hope.

Blessings and curses

Prayers are not always benign. Nor are gods and goddesses. They can destroy as well as create. In Tibetan Buddhism, wrathful deities appear as demons in order to help sentient beings towards enlightenment, but the same deities are also enlightened beings who in other circumstances appear in peaceful forms. The Hindu god Shiva destroys as well as creates.

The God of the Abrahamic religions has the power to curse as well as to bless, and exercised this power most spectacularly in the ten plagues on the land and people of Egypt, which finally persuaded the Pharaoh to let the Jewish people go, to begin their journey through the wilderness to the Promised Land. The last of these terrible plagues was the killing of the first-born sons of the Egyptians and of their cattle. In Islam, among his ninety-nine names, Allah is both 'The Giver of Life' and 'The Bringer of Death'; both 'The Benefactor' and 'The Afflicter'.

Those who have the power to bless have the power to curse. When I lived in India, one evening when I was talking to a Hindu friend in a street, I saw a pair of *sadhus*, holy men with their long hair, flowing orange robes and beards begging from house to house. Most people gave them something, and in return received a blessing. I commented to my Hindu friend about the high level of generosity that these holy men evoked. 'Yes,' he replied, 'People want the sadhus' blessings and are afraid of being cursed if they do not give.'

In the medical context, blessings and curses are mirrored by placebos and nocebos. The placebo effect occurs when people are hopeful and are being cared for. If they are in a double-blind clinical trial, where neither the patients nor the doctors know who is getting the new wonder drug and who is getting the blank pills, the placebo often has remarkably positive effects, because people who are not receiving it hope that they might be, and that they will get better. Many potential drugs fail in clinical trials, because the placebo effect works as well as the drug itself.

Nocebos are negative placebos, when people expect negative effects. If they believe they have drunk a poison, their health may become worse. In the most extreme cases, people who believe they have been cursed to death actually die. In his classic paper on voodoo deaths, the physiologist Walter Cannon outlined three essential elements that made these curses effective, and usually fatal within a few days.

First the victims, their family members and acquaintances need to believe in the power of the curse to cause death. Second, all previous known victims must have died, unless the curse was removed by the hexer or by another witch doctor. Third, everyone must treat the victim as if he will die, leaving him isolated and alone.[53]

Some critics of conventional medicine claim that official diagnoses of potentially terminal diseases can act as medical curses, causing voodoo deaths. Those who believe themselves doomed to die, and whose families believe they are doomed, are indeed more likely to die. This presents a serious problem for doctors, who have a duty to inform patients of their medical condition, and tell them about the risks of medical procedures.

But patients find some comments from doctors very frightening. Some of the most common include: 'You are living on borrowed time,' 'It can only get worse,' 'You have six months to live,' or 'You have a time bomb in your chest.' The practices of medicine often include placebo responses, but they include nocebo responses as well.[54]

Curses are not confined to witch doctors, wandering sadhus and insensitive doctors. They are part of mainstream religion, too. In the *Book of Common Prayer* of the Church of England, first compiled in 1549, there is a series of curses to be made on Ash Wednesday, the first day of Lent, called a Commination. These include, 'Cursed is he that curseth his father or mother. Cursed is he that maketh the blind to go out of his way. Cursed is he that taketh reward to slay the innocent.' Today these curses are rarely if ever used, but the fact that they are still part of the official prayer book of the Church of England underlines the fact that blessings and curses are closely related. The power to do good and the power to do harm both work through intentions.

The English word 'blessing', from the Old English word *bloedsian*, has connotations that the equivalent Latin-based word 'benediction' does not have. In Latin, *benedictio* is about well (*bene*) speaking (*dictio*), whereas *bloedsian* means to make holy through blood (*bloed*), in other words to consecrate through a blood sacrifice – and to make happy by rendering holy.[55]

In the Bible, to be blessed means to be favoured by God. Blessings come from God, and to pray for a blessing on someone or something is to pray for God's favour and protection. The Priestly Blessing in the Old Testament, first transmitted by Moses, was pronounced by priests in the Temple in Jerusalem, and is still widely used in synagogues and in churches today:

> *The Lord bless you and keep you;*
> *The Lord make his face to shine upon you and be gracious to*
>   *you;*
> *The Lord lift up his countenance upon you and give you peace.*
>
>   (Numbers 6: 24–6)[56]

In Islam, too, Allah is the source of all blessings, and Muslims are enjoined continually to give thanks for the many blessings they receive. If they are thankless they will be condemned. As the Koran puts it, 'And remember when your Lord proclaimed, "If you are grateful, I will surely increase you [in favour], but if you are thankless, My punishment is severe."' (Surah 14: 7).

In Judaism, Christianity and Islam, many greeting and parting ritual phrases are blessings, as in 'God bless you', and 'goodbye', shortened from 'God be with you', and the Muslim *as-salāmu alaikum wa rahmatul-lāhi wa barakātuh*, meaning 'may peace, mercy and the blessings of God be upon you.'

Blessings are a form of prayer in which all of us can participate, both by using traditional forms of words, or making our own.

The Irish poet and priest, John O'Donohue (1956–2008), invited us to rediscover this power that can flow through us all:

In the parched deserts of post-modernity, a blessing can be like the discovery of a fresh well. It would be lovely if we could rediscover our power to bless each other. I believe each of us can bless. When a blessing is invoked, it changes the atmosphere. Some of the plenitude flows into our hearts from the invisible neighbourhood of loving kindness. In the light and reverence of blessing, a person or situation becomes illuminated in a completely new way . . . The language of blessing is an invocation, a calling forth . . . In the invocation of blessings, the word 'may' is the spring through which the Holy Spirit is invoked to surge into presence and effect. The Holy Spirit is the subtle presence and secret energy behind every blessing.[57]

In his book, *Benedictus: A Book of Blessings*, O'Donohue offers a great range of blessings including blessings for fire, for the morning, for courage, for new fathers and mothers, for people at thresholds of changes in their life, before and after meals, for celebrations, for those who are suffering, and for the dying. One of the blessings is for a new home:

May this house shelter your life
When you come in home here.
May all the weight of the world
Fall from your shoulders.

May your heart be tranquil here,
Blessed by peace the world cannot give.

May this home be a lucky place,
Where the graces your life desires
Always find the pathway to your door . . .[58]

This blessing is a reminder of the fact that all over the world in traditional societies, people want their house to be blessed when they move in. When I lived in the ashram of Father Bede Griffiths in South India, every few weeks a Hindu or Christian family from the nearby village would ask him to come to bless their new home, which he did with prayers and holy water. I sometimes went with him.

In many Christian countries, it is traditional for people to ask priests or pastors to bless their new house, which they do by walking through all the rooms saying prayers and sprinkling holy water. My wife and I asked our local Anglican priest to bless our home in London, which he did very graciously. Father Bede also blessed it when he was visiting London.

Even in the modern secular world, many people recognise a need for some kind of ceremony when they move to a new home. Housewarming parties are the secular descendants of traditional blessing ceremonies found in cultures all over the world. In the original housewarmings, guests would bring firewood to create fires in all the house's fireplaces, to create warmth and light and to drive out harmful spirits. Today many people bring gifts as a form of blessing. In the Jewish tradition, traditional offerings include bread, so that members of the household do not go hungry, salt for savour, and sugar so that the lives will be full of sweetness.

Blessings are a form of prayer that almost everyone can appreciate and value.

Two practices of prayer

TRY PRAYING

If you pray regularly, then you already have a channel through which you can ask for guidance about praying.

If you pray only irregularly or used to pray and you have stopped, I suggest making it a regular practice, ideally for at least fifteen minutes a day, and preferably at a more or less fixed time, for instance, before you go to work or before you go to bed.[59]

Unless you have a strong objection to your own ancestral tradition, it is best to pray in that tradition. You will feel more at home, more spiritually comfortable. If you are a Muslim, pray Muslim prayers; if a Hindu, Hindu prayers; a Jew, Jewish prayers; a Christian, Christian prayers. And it is best to start with prayers that are central to your ancestral tradition, for example, the Lord's Prayer for people from a Christian background.

Prayers start with an invocation to the spiritual being to whom you are praying, as in 'Our Father who art in heaven', 'Hail Mary, full of grace', or 'Om Namah Shivaya' or 'O Allah'. For Hindus there are many gods or goddesses to whom they can address their prayers. Some Protestant Christians address their prayers only to God the Father, but Roman Catholics and Anglicans address their prayers to God in any of his three aspects or modes, as Father, as in 'Our Father, who art in heaven,' as the Son, as in the Jesus prayer 'Lord Jesus Christ have mercy on me', and as the Holy Spirit, as in 'Come Holy Spirit' – or as all three together, as the Holy Trinity.

As well as prayers with a set form, try expressing your thanks for what has happened since you last prayed in your own words. Ask for forgiveness in your own words, express your doubts and fears in your own words, and ask for help, guidance, protection and blessings in your own words, for others and for yourself. End the prayer with a mantra that is central to your own tradition, such as *Aum* in Hinduism and Buddhism, and *Ameen* or *Amen* in the Jewish, Christian and Muslim traditions.

TRY PRAYING TOGETHER

It is a powerfully bonding experience to pray with other people, and for a community to pray together. The easiest way to do this is to attend a religious ceremony in your own tradition; for Jews at a synagogue, for Muslims a mosque, for Christians a church, for Hindus and Sikhs a temple. There you will be led in prayers, and you will say or chant prayers together.

Ways to Go Beyond and Why They Work

In the secular world, it is hard to pray with colleagues at work or in other social groups, because these groups usually include people from different religious traditions and people who are non-religious, or anti-religious. In some families, praying together is still possible, and is most easily carried out by giving thanks together before meals with graces, said or sung.

6

Holy Days and Festivals

Holy days are literally holidays. They are days on which normal work comes to a stop so that the community can take part in collective celebrations.

The earliest evidence of ceremonial activities comes from caves. The oldest known work of art, more than 40,000 years old, is a sculpture of a human figure with a lion's head, the Lion Man, made from mammoth ivory. The Lion Man was found in a dark inner chamber in a north-facing cave in Germany, which was probably not inhabited, but used for special gatherings.[1]

In other caves used for ceremonies in Ice Age Europe, remarkable paintings on the walls show a range of animal species, like bison and reindeer, as well as human figures with animal heads. Some of these paintings date back more than 35,000 years. Similar animal-human figures are found in ancient cave paintings in southern Africa. No one knows why these paintings were made, or what altered states of consciousness may have given rise to these visions of animal-headed humans.[2]

But it seems likely that these paintings were made at ritual centres, often deep within caves, where people gathered for ceremonies at which they sang and danced. They could not have been going about their normal hunting and gathering activities while they were participating in rituals deep in caves. These ceremonies must have taken place on special days, on holy days.

Like these remote ancestors, recent hunter-gatherers had seasonal ceremonies related to success in hunting and to the fertility of the animals they hunted, to ensure abundance. Among the

Eveny peoples of Siberia, nomadic reindeer herders who live in the coldest inhabited place on earth, in living memory, there was a traditional festival on Midsummer Day:

> During the white night of the Arctic summer, a rope was stretched between two large trees to represent a gateway to the sky. As the sun rose above the horizon in the early dawn, this gateway was filled with the purifying smoke of the aromatic mountain rhododendron, which drifted over the area from two separate bonfires. Each person passed around the first fire anticlockwise, against the direction of the sun, to symbolise the death of the old year and the burning away of its illnesses. They then moved around the second fire in a clockwise direction, following the sun's own motion to symbolise the birth of the new year.
>
> It was at this moment, while elders prayed to the sun for success in hunting, an increase in reindeer, strong sons and beautiful daughters, that each person was said to be borne aloft on the back of a reindeer, which carried its human passenger towards a land of happiness and plenty near the sun. There they received a blessing, salvation, and renewal . . . This rite was followed by a *hedje*, a circle dance in the direction of the sun, and a feast of plenty.[3]

Seasonal rituals were probably celebrated by hunter-gatherer societies for tens of thousands of years before the development of agriculture. When people began to cultivate crops and domesticate animals, a new sequence of seasonal rituals for fertility, rain, abundance and thanksgiving evolved.[4] The timings of these festivals varied from place to place according to the climate. The main practical concerns of these festivals were the most basic of human needs and desires, food and children.[5]

All these festivals necessitated the stopping of work. In addition, in many agricultural societies, more or less regular rest days were also devoted to markets.[6] Those who attended them had to abandon for the time being their usual occupations, and these days also gave opportunities for social activities, sports and amusements.

The seven-day week that we now take for granted was first instituted by the Babylonians, who named the days after the seven heavenly bodies visible to the naked eye: the sun, moon, Mars, Mercury, Jupiter, Venus and Saturn.[7] The Jewish seven-day week is probably derived from the Babylonian, and the creation story in the book of Genesis portrays God as creating the heavens, the earth and everything in them over a six-day period. He rested on the seventh. The sanctification of the seventh day as a day of rest, the Sabbath, became one of the most distinctive features of Judaism.

Christians adopted the principle of the Sabbath, but Sunday was their holy day rather than Saturday. Muslims, distinguishing themselves from Christians and Jews, made Friday their day of assembly, when all could pray together. Weekly rest days are now built into the calendars of all nations on earth.

In this chapter, I start with the Jewish Sabbath, or *Shabbat*, and the related holy days of Christianity and Islam, and look at the erosion of these traditions through secularisation and the 24/7 culture. I explore ways in which we can rediscover these days and experience their healing power. I then discuss seasonal festivals, principally those in the Christian tradition, which are still a major part of the calendar in all historically Christian countries.

The Jewish Shabbat

For Jewish people, the *Shabbat* or Sabbath was and still is regarded as a symbol of the close relationship between the people of Israel and God. It was holy because it was a sign of the chosen people's sanctification by God, and at the same time separated the people of Israel from the rest of the world. In the words of a popular

Jewish table hymn, dating from the twelfth century AD, 'I keep the Sabbath, God keeps me: It is an eternal sign between Him and me.'[8]

The fourth of the Ten Commandments defines this holy day as a time of rest from work. The very word *Shabbat* means rest:

> Remember the Sabbath day and keep it holy. Six days you shall labour and do all your work. But the seventh day is a sabbath to the Lord your God; you shall not do any work – you, your son or your daughter, your male or female slave, your livestock, or the alien resident in your towns. For in six days the Lord made heaven and earth, the sea, and all that is in them, but rested the seventh day; therefore the Lord blessed the sabbath day and consecrated it. (Exodus 20: 8–11)[9]

For Jewish people, Shabbat begins at sundown on Friday and ends at sundown on Saturday. To this day, many Jewish families celebrate Shabbat by eating a meal together on Friday evening. To start with, a woman of the family lights the candles, of which there should be at least two, representing the dual commandments to remember and to keep the sabbath. The woman then says a prayer of blessing: 'Blessed are you, Lord, our God, sovereign of the universe, who has sanctified us with his commandments and commanded us to light the lights of Shabbat. Ameen.'

Shabbat customarily includes three celebratory meals: dinner on Friday evening, lunch on Saturday and a third meal in the late afternoon on Saturday. Many Jews attend one or more synagogue services on Shabbat, which are held on Friday evening, Saturday morning and late Saturday afternoon. Many Jews also wear festive clothing for Shabbat. The Sabbath ends on Saturday evening with a ceremony called Havdalah, meaning 'separation', which involves lighting a special candle with seven wicks, blessing a cup of wine and smelling sweet spices.

For Orthodox Jews, there are many restrictions on what they can do on Shabbat. No fewer than thirty-nine categories of work are prohibited, including shearing wool, slaughtering animals, kindling or extinguishing fires, and transporting objects between private and public domains. Orthodox Jews extend these categories to modern life, and include the switching on or off of electrical devices, because they fall under the category of kindling or extinguishing a fire. There are even lifts that work in 'Sabbath mode', stopping automatically at every floor, enabling people to step on and off without having to press buttons.

For non-Orthodox Jews, the observation of the Sabbath involves fewer restrictions, and disputes about what is permitted and not permitted have gone on for millennia. Jesus had several disagreements with conservative Jews and took a decidedly liberal view, saying, 'The Sabbath was made for human kind, and not human kind for the Sabbath.' (Mark 2: 27).[10] There has been a continual tension between legalistic regulations designed to preserve the sanctity of the Sabbath, and the spirit of the Sabbath, which is to provide a work-free opportunity for family life, the worship of God, study, recreation, music, delight, and making love.

Over the centuries, many critics of Judaism have objected to this weekly observance. The classical Roman author Seneca (c. 4 BC–AD 65) made a point that secularists make today, seeing the Sabbath as an obstacle to productive work: 'To remain idle every seventh day is to lose a seventh part of life, while many pressing interests suffer by this idleness.'

Rabbi Abraham Heschel (1907–72), one of the great Jewish teachers of the twentieth century, emphasised that, 'Judaism is a *religion of time* aiming at the *sanctification of time*.' One of his metaphors was of the Sabbath as an island of stillness in the tempestuous ocean of time and toil: 'The island is the seventh day, the Sabbath, a day of detachment from things, instruments and practical affairs as well as of attachment to the spirit.'

He argued that the whole point of the observation of the Sabbath was to celebrate time rather than space. After the destruction of the temple in Jerusalem by the Romans in AD 70, the Jewish people were living in exile. While other people still had their holy places, temples, shrines, cathedrals and churches, the Jewish people focused on holy times rather than holy places.

'Six days a week we live under the tyranny of things of space; on the Sabbath we try to become attuned to *holiness in time*. It is a day on which we are called upon to share what is eternal in time, to turn from the results of creation to the mystery of creation; from the world of creation to the creation of the world.'[11]

Heschel saw the other days of the week as leading up to the Sabbath: 'Judaism tries to foster the vision of life as a pilgrimage to the seventh day; belonging to the Sabbath all days of the week, which is a form of longing for the eternal Sabbath all the days of our lives.'[12]

Sundays and weekends

The first Christians, Jesus's immediate disciples, were Jews who continued to observe the Sabbath, but they also gathered together on Sunday, the first day of the week, to celebrate the central event in the Christian faith, the Resurrection of Jesus, which happened on Easter Sunday. As Christianity spread among non-Jewish people, Sunday came to replace the Sabbath as the day of rest. The early Christians called Sunday the Day of the Lord, *Die Dominicus* in Latin, from which the words for Sunday in Latin-based languages are derived: *Domenica* in Italian, *Domingo* in Spanish, *Dimanche* in French. In northern Europe, the older name persisted, as in *Sunday* in English, *Sonntag* in German, and *Söndag* in Swedish.

The Roman Emperor Constantine enacted the first official recognition of Sunday as a legal day of rest in AD 321: all courts of law, inhabitants of towns and workshops were to be at rest on Sunday, which was still called the Day of the Sun, *Dies Solis*. Subsequent Roman emperors went further, and banned theatres and circuses on Sunday.

In England, as early as the seventh century, the laws of Saxon kings prohibited work on Sunday. Legislation restricting the opening of shops and a range of other commercial activities continued until the late twentieth century, and some restrictions on large stores remain even today. And almost all government offices, businesses, banks, schools and universities are closed on Sundays.

In civil law, Sunday is defined as running from midnight on Saturday night to midnight on Sunday night, but ecclesiastical law follows the Jewish custom of starting the day at sunset. Thus many Roman Catholics attend Sunday mass on Saturday evenings.

In the United Kingdom, as in many other countries, there is an ongoing tension between keeping Sunday as a day of rest, and secular and commercial forces trying to sweep away restrictions.

This conflict has a long history. In the nineteenth century, strict Sabbatarians, like the Lord's Day Observance Society, established in 1834, tried to prevent sporting activities and the consumption of alcohol on Sundays. Anti-religious organisations, like the Sunday League, tried to promote all kinds of activities on Sundays, as long as they were not religious. But there were many intermediate positions, including proponents of muscular Christianity, who promoted playing football and other games on Sundays, although not at service times. Churches founded many football clubs, including Manchester City and Celtic. Sunday football leagues persist to this day.

The practice of keeping Sunday as a work-free day spread throughout the world in the Spanish, Portuguese, British, French, Dutch and other European empires. Banks, businesses and government offices were closed. This practice still continues, even in countries like India and China, where Christians are a small minority.

Starting in the mid-nineteenth century, skilled workers in British factories were given Saturday afternoons off. By the early twentieth century, a holiday on Saturday was widespread. The weekend was born. It is now standard practice for banks, schools and offices to close all day Saturday, though this does not apply to shops, restaurants, entertainment venues and transport systems.

The conflict experienced by early Christians between the observation of the Sabbath on Saturdays and the celebration of Sunday as the Lord's Day has been resolved. Both days are now holidays or holy days. Thus the Judeo-Christian week has spread worldwide, with the exception of some Muslim countries, where there is a Friday–Saturday weekend or a Thursday–Friday weekend.

For Muslims, Friday is above all a day on which believers gather together to pray in congregation, the day of assembly. Praying together on Fridays is one of the fundamental duties of Muslims. In the words of the Koran: 'O you who believe! When the call to prayer is proclaimed on Friday hasten earnestly to the remembrance of God, and leave aside business.' (Surah 62: 19.) Although businesses should close at the time of congregational prayer, they are not prohibited from opening at other times on Fridays, and restrictions on work are less stringent in Islam than in traditional Jewish and Christian communities.

In modern secular societies, there are continual attempts to sweep aside all restrictions and fully commercialise the day of rest. For example, in the United Kingdom there was a major political debate leading up to the 1994 Sunday Trading Act, which allowed shops to open on Sundays, but restricted the opening hours of larger stores. This removal of most prohibitions on Sunday trading was opposed by the Lord's Day Observance Society and the Keep Sunday Special campaign, a coalition group that included the shopworkers' trade union. This union finally withdrew its opposition in return for a promise that working on Sundays would be strictly voluntary.[13]

Many people prefer not to work on a Sunday, even if they are paid more and compensated with rest days on other days of the week, because there is less scope for them to enjoy these alternative rest days with their families and friends. Family and communal activities depend on shared holidays.

The 1994 debate took into account the needs of workers. By contrast, the 24/7 culture of the internet extinguishes almost all distinctions of time, with no debate at all. All time is commercial time. Google, Facebook, Amazon and the entire internet function continuously, with no rest periods, and no holy days. Until recently, at least in Britain, Sundays were days on which there were no postal or courier deliveries. But now Amazon and several other delivery companies have made Sundays like any other day. I myself now avoid ordering anything on Fridays or Saturdays because I do not want deliveries on Sundays.

We are in a paradoxical situation. Thanks to religious traditions, social reforms, and campaigns for workers' rights, most people now have two days of rest a week. But the workaholic 24/7 culture threatens to obliterate this distinction so that every day is a shopping day, a working day, an online-banking day, and a social-media day.

Although there is officially more leisure time than ever before, many people feel overwhelmed by the amount of work they have to do, and are exhausted by it. One study of American work patterns concluded, 'Time has become a precious commodity and the ultimate scarcity for millions of Americans.' A survey by the *Wall Street Journal* found that forty per cent of Americans said that lack of time was a bigger problem for them than lack of money. And as the American work ethic spreads around the world, this is becoming a global problem.[14]

Traditional cultures recognised the need for people – and

working animals – to have days of rest. Not surprisingly, modern research bears out this traditional understanding. For instance, an international survey in 2016 showed that people who felt they did not get enough rest scored much lower on wellbeing scales.[15] A wide range of scientific studies has now shown how beneficial periods of rest can be.

As a review article in *Scientific American* put it, 'Downtime replenishes the brain's stores of attention and motivation, encourages productivity and creativity, and is essential to both achieve our highest levels of performance and simply form stable memories in everyday life . . . Moments of respite may even be necessary to keep one's moral compass and maintain a sense of self.'[16]

In the US, a new genre of medically based self-help books advocate the benefits of a 'stop day', or a '24/6' lifestyle, so that one day a week is set aside from work.[17] The day of rest is being reinvented on a scientific rather than a religious basis. But lest this be thought to be advocating mere idleness, the new slogan is 'rest for success.'[18]

Seasonal festivals

All over the earth, except in the humid equatorial zones, there are unmistakable markers of the seasons. In the northern hemisphere, at the summer solstice the sun rises in the north-east at its utmost northern limit on the horizon, and sets at its ultimate point on the horizon in the north-west. Thereafter, it rises further south along the horizon every day until the winter solstice; it then rises at its southernmost point on the eastern horizon and sets at its southernmost point on the western horizon. At the solstices, the sun seems to stop for about three days in this position; indeed, the word solstice means sun-stoppage (Latin: *sol* = sun; *stitium* = stoppage). To know when the solstices are happening, all that is needed is to observe the horizon from a particular place.

Another way is to look at the shadows cast by a pole or standing stone. Before the summer solstice, the shadows cast by the rising sun day by day move gradually towards the south; and before the winter solstice, they move gradually towards the north. Then they stay in the same place for about three days; thereafter the direction of their movement reverses.

At the equinoxes, when the night and day are of equal length, the sun rises on the horizon due east and sets due west.

Some ancient stone circles, like Stonehenge in England, seem to have been designed to indicate the direction of the rising sun at the summer solstice, and some ancient passage tombs, like Newgrange in Ireland, dating from about 3,000 BC, are aligned with the sunrise on the winter solstice, when the sun shines directly along the long passage, lighting up the inner chamber. No one knows what kind of ceremonies were celebrated at these great Neolithic monuments, but it seems very likely that they were important centres at which people gathered for seasonal festivals. The enormous effort involved in constructing them shows how important they were.

The solstices and equinoxes provide a natural way of dividing the year into quarters, and in many cultures the main festivals occur around these divisions of the year. In medieval Europe, the festivals connected with the cardinal points of the solar year were Christmas on 25 December, just after the winter solstice; the feast of the Annunciation of the Blessed Virgin Mary or Lady Day on 25 March, shortly after the spring equinox; midsummer day or St John's day on 24 June, just after the summer solstice; and Michaelmas, the feast of St Michael and All Angels on 29 September, soon after the autumn equinox. In England these days were called quarter days, when quarterly rents and other dues were paid.

The medieval Christian calendar had a second set of quarter days between the solar cardinal points, sometimes called cross-quarter

days: Candlemas on 2 February; May Day on 1 May; Lammas on
1 August; and All Saints' Day on 1 November. In the pre-Christian
period in Ireland and in parts of Scotland, these cross-quarter days
were important times of festivals: the ancient Irish feast of Imbolc
was on 1 February, and was later rededicated to St Brigid; Beltane
was on 1 May; Lughnasadh, Christianised as Lammas, was on
1 August; and Samhain, which became the feast of All Saints or All
Hallows, was on 1 November.

In many cultures all over the world, festivals and ceremonies
mark these annual changes. The details vary, depending on the
climate. In northern Europe, the spring equinox is a time of
regrowth, blossoming and preparation for the long, fertile days of
the summer, with harvests in the autumn. In most of India, rain-fed
crops are sown when the monsoon begins in June or July. Many
of these *kharif* crops are harvested around October; often second
crops are then planted and harvested in February or March at the
end of the *rabi* season. And, of course, in the southern hemisphere
the seasons are opposite.

As well as the festivals I discuss below, there are many national
festivals and rituals, like the American Thanksgiving in November,
and also many local festivals, like the patronal festivals of parish
churches and cathedrals in Catholic countries, which occur on the
feast day of the patron saint.

The winter solstice and Christmas

All over Europe there were festivals around the time of the winter
solstice. In northern Europe, the midwinter festival of Yule was
celebrated with fires and light. In ancient Rome, the winter festival
was called the Saturnalia, beginning on 17 December, and lasting
for a week. The law courts and the schools were closed, and all
commercial activities, legal business and household chores ceased.

The time was given up to feasting, gambling and the reversal of the established order of society. Slaves were served by their masters and sat at table with them, and were able to insult them with impunity. People exchanged presents, including candles.[19] Immediately afterwards, 25 December was a sun festival celebrating the birth of the sun god, *Sol invictus*, the invincible sun, and was especially important for followers of the Mithraic cult.[20]

No one knew Christ's actual birthday, but in a calendar drawn up in Rome in AD 336, 25 December was set apart for the celebration of his birthday, and Christmas took on many of the qualities of the pre-existing mid-winter festivals like Saturnalia, and still does to this day.[21] The Roman New Year feast took place from 1 to 3 January.

In AD 567, the Council of Tours declared that the whole period of twelve days between the feast of the birth of Christ on 25 December and the feast of the Epiphany on 6 January formed one continuous cycle of festivities, the twelve days of Christmas.[22] In England in the Middle Ages, these twelve days were a time for feasting and for hospitality, when the rich were expected to share their good fortune with the poor. There were also plays, singing and music, as well as more unruly activities such as 'disguising' or 'mumming', in which some people wore masks. Cross-dressing was common.

The Protestant Reformers disapproved, and the Puritan radicals tried to abolish Christmas altogether. But seasonal festivals are deeply embedded in human cultures. The Puritans met with little success in England, because Christmas was just too popular. In Scotland, Protestantism was more severe and the suppression of Christmas more successful. Christmas did not become an official holiday in Scotland until 1958![23] The result was that the mid-winter celebrations were displaced to Hogmanay, on New Year's Eve.

The fortunes of Christmas have waxed and waned. In the early nineteenth century, many people in England saw Christmas as an anachronism, and for factory owners it was an unwelcome interruption of profitable work.²⁴ But attempts to turn the Christmas season into business as usual was strongly opposed by many workers in traditional industries. By the mid-nineteenth century, a large-scale revival of Christmas was under way, encouraged by churches, social reformers and writers such as Charles Dickens.

The details of Christmas celebrations vary from country to country. In some, the principal celebration is on Christmas Eve, when people give presents to each other and Father Christmas brings presents for children; in others, Christmas Day itself is the main day for feasting and present-giving. And in some southern European countries, as in Italy and Spain, present-giving happens on the feast of the Epiphany, 6 January, when the Three Wise Men gave their presents to the baby Jesus.

In seasonal festivals there are many variations from place to place. The details often seem arbitrary, and they depend to a large degree on local customs. But in a given locality, there is a general agreement on the date and ways of celebrating, because these are communal festivals that involve people of all ages, bringing people together in families, and families into communities. In order to work, there needs to be a general consensus, even if it varies from place to place.

Spring festivals and Easter

In many cultures there are festivals around the time of the vernal equinox. In some cases, as in northern Europe, they are associated with themes of new life as the growing season begins. In other places, as in the Middle East and India, they are associated with the harvest of crops that have grown through the winter season.

In India, the festival of Holi, at full moon in late February or early March, is both a harvest festival and a celebration of fertility. The first evening of Holi involves a reversal of social roles, as in the Roman Saturnalia, as I discussed in Chapter Four. The following day is a festival of colours, in which people often take cannabis-containing drinks and drench each other with coloured water or powders. There is much singing and dancing.

Traditional Holi celebrations were often explicitly erotic. A description of this festival written by an Englishman in the early twentieth century commented disapprovingly, 'The orgies of obscenity which welcome the return of spring are scarcely veiled . . . The law practically permits of any excess, the god encourages it, and the nature of the people . . . revels in its own unbridled enjoyment of indecency. Street dances, bonfires and the throwing of red and yellow powder upon the passers-by remind the Occidental visitor of a Western carnival; but no Western carnival at its worst is so frankly sensual as is the spring-festival of India.'[25]

The God who revealed himself to Moses and who guided the Jewish people on their journey through wilderness to the Promised Land was a desert god. But his character changed after the people settled in Palestine, and Yahweh took over the role and functions of the indigenous vegetation gods. The agricultural festivals were reinterpreted in terms of Jewish history. As Rabbi Heschel put it:

The festivals of ancient peoples were intimately linked with nature's seasons. They celebrated what happened in the life of nature in the respective seasons . . . In Judaism, Passover, origi- nally a spring festival, became a celebration of the Exodus from Egypt; the Feast of Weeks, an old harvest festival at the end of the wheat harvest (Exodus 23:16; 34:22), became the celebration of the day on which the Torah was given at Sinai; the Feast of the Booths, an old festival of vintage (Exodus 23:16), commem- orates the dwelling of the Israelites in booths during their sojourn in the wilderness (Leviticus 23: 42–3).

To Israel the unique events of historic time were spiritually more significant than the repetitive processes in the cycle of nature, even though physical sustenance depended on the latter. While the deities of other peoples were associated with places or things, the god of Israel was the god of events: the Redeemer from slavery, the Revealer of the Torah, manifesting Himself in events of history rather than in things or places.[26]

Christianity inherited from Judaism this historical transformation of seasonal festivals. Jesus' last supper with his disciples took place at Passover, on the eve of his crucifixion. His death on the cross on Good Friday and his resurrection on Easter Sunday are the central events that Christians commemorate in the festival of Easter. Like Passover, the timing of this festival depends on the full moon and the vernal equinox, but whereas Passover happens on the first full moon after the vernal equinox, the celebration of the Resurrection has to be on a Sunday.

Thus, Easter Sunday is the first Sunday after the first full moon after the vernal equinox. The date of Easter Sunday can be as early as 22 March and as late as 25 April. Because Easter is a moveable feast, so are all the feast days associated with it. The fasting period of Lent begins forty-six days before Easter Sunday, on Ash Wednesday, which can be as early as 4 February or as late as 10 March. The day before Ash Wednesday is Shrove Tuesday, or Carnival, the last chance for feasting, singing and dancing before Easter, celebrated most spectacularly in Brazil. The feast of the Ascension, when Christ's resurrected body rose into the sky, is forty days after Easter, and Pentecost, or Whitsun, the festival of the Holy Spirit falls on the seventh Sunday after Easter Sunday.

Consequently Easter has several aspects or levels. It is a spring festival, a time of regeneration and rebirth, with presents of eggs and images of rabbits – which breed like rabbits. As a commemoration of Jesus' death and resurrection, it also inherits archetypal images of dying and resurrected gods, like the ancient Egyptian god Osiris, a god of death and resurrection, and of sprouting vegetation.[27]

All around the eastern Mediterranean there were annual cele-brations of the death and resurrection of a god associated with crops. As the anthropologist James Frazer put it, 'Under the names of Osiris, Tammuz, Adonis and Attis, the peoples of Egypt and Western Asia represented the yearly decay and revival of life, especially of vegetable life, which they personified as a god who annually died and rose again from the dead.'[28]

In his book, *The Golden Bough*, Frazer pointed out that in many cultures the first fruits, the newly harvested crops, were eaten sacramentally as the body of the vegetation spirit. In a chapter called 'Eating the God', he compiled numerous examples from all around the world.[29] Frazer also drew attention to many instances of sacrificial kingship, in which kings, who were believed to be endowed with divine powers, were violently sacrificed so that they did not grow old and feeble, and endanger the life of the group.[30]

Yet another element in the Easter story is that of the sacrificial animal whose death ensures the safety of others. In the context of the Jewish tradition, this took place in three ways (discussed in *SSP*, chapter 5). First, the sacrifice of a ram by Abraham instead of his own son involved a substitution of animal for human sacri-fice. Second, the sacrifice of a lamb at Passover protected the people of Israel from the death and destruction visited on the Egyptians. And third, in an annual Jewish ceremony, the sins of the people were laid upon a goat, the scapegoat, which was driven away into the wilderness, where it perished, taking away the sins of the people with it.

The death of Jesus on the cross was like that of a sacrificial animal, taking away the sins of the world. He was the Lamb of God. The old pattern was reversed: instead of an animal being substituted for a human sacrifice, a human was substituted for an animal in this full and final sacrifice.

Frazer believed in an evolutionary ideology that saw primitive humanity as engaged in magical practices, which gradually gave way to religious belief, and then finally reached the highest level of human development in scientific thought. When I was a teenager, one of my science teachers introduced me to Frazer's ideas, which had their intended effect. I recognised that there were many elements in the Christian religion that were rooted in ancient mythologies, pre-Christian fertility cults and magical rituals. These seemed like good reasons for rejecting Christianity in favour of science and reason.

I am still in favour of science and reason, though no longer a believer in the ideologies of scientism and rationalism. But as a practising Christian, and as a participant in the Christian festivals and sacraments, I now see these deep archetypal elements as a strength, not a weakness. The continuity of the seasonal festivals of Christianity with pre-Christian festivals and myths makes them more powerful, not less powerful, and gives them greater meaning and significance.

May Day

May is the month of the goddess, called after the Roman goddess Maia. In the Roman Catholic Church, May is called the month of Mary.

May Day, 1 May, was a major feast of vegetation, new life and fertility all over northern Europe. In England, groups of people or couples went out into the woods in the early morning to gather flowers and greenery to decorate their houses. There are many descriptions in medieval English literature of the 'bringing in of May'. In Chaucer's *Court of Love*, 'Forth goeth all the Court both most and least/ To fetch the flowers fresh and branch and bloom.' His heroine Emelie goes out at sunrise 'to do May observance' by gathering 'flowers pretty white and red/ To make a subtle garland for her head.'[31]

A writer in the sixteenth century described how Londoners

would try to spend May morning in 'the sweet meadows and green woods, there to rejoice their spirits with the beauty and savour of sweet flowers and with the harmony of birds.'[32]

Inevitably the Puritans disapproved and there were many denunciations of the celebration of May Day in the sixteenth and seventeenth centuries. The traditional practices were deemed to be frivolous – or worse, a licence for debauchery. Nevertheless, the idea of making love in the woods may have been more exciting than the reality. The weather would often have been chilly and the woods damp.

In the late twentieth century, demographic historians carried out patient research to investigate whether there was in fact a surge of conceptions at this festival, based on birth rates nine months later. They found no such boom. There was, however, a surge of conceptions later in the summer when the weather was warmer.[33]

While out at dawn on May Day itself, girls and women washed their faces with dew to enhance their complexions; many believed that kissing the dew would make them beautiful.[34] In some parts of England, May queens were enthroned and garlanded. In other places, both a king and queen, a young man and a young woman, were elected by the youths, and helped lead the celebrations.[35] In many places people sang and danced around a maypole, decorated with flowers and ribbons.

In Ireland and parts of Scotland and Wales, this festival was called Beltane and took a different form. Following a druidic custom, people made pairs of bonfires, and drove their cattle between them. Beltane coincided with the moving of cattle to summer pastures, and driving them between the fires was believed to protect them from diseases and other misfortunes. These fires also blessed the humans who leapt over them.

Sir William Wilde, Oscar Wilde's father, published a description of these Irish customs in 1852:

With some, particularly the younger portion, this was a mere diversion, to which they attached no particular meaning, yet others performed it with a deeper intention, and evidently as a religious rite. Thus, many of the old people might be seen circumambulating the fire, and repeating to themselves certain prayers.

If a man was about to perform a long journey, he leaped backwards and forwards three times through the fire, to give him success in his undertaking. If about to wed, he did it to purify himself for the marriage state. If going to undertake some hazardous enterprise, he passed through the fire to render himself invulnerable.

As the fire sunk low, the girls tripped across it to procure good husbands; women great with child might be seen stepping through it to ensure a happy delivery, and children were also carried across the smouldering ashes. At the end, the embers were thrown among the sprouting crops to protect them, while each household carried some back to kindle a new fire in its hearth.[36]

In Germany and several nearby countries, the eve of May Day, called Walpurgis night, was believed to be the time that witches met on the Brocken, the highest peak in the Harz Mountains. The night was celebrated with dancing and the burning of bonfires to drive away the witches.

In 1889, the Socialist International adopted May Day as a day of socialist celebration.[37] In the Soviet Union, it was the occasion for vast military parades through Red Square in Moscow. In Britain, the Labour government instituted a new public holiday for May Day in 1978 amidst intense controversy, because of its association with socialism and the Soviet Union.[38] When the Conservatives returned to power the next year, they planned to scrap this new holiday and replace it with a more patriotic holiday for Trafalgar Day in October. The response to this proposal was lukewarm. Most people preferred a day off in early summer to one in late autumn, and the plan was quietly shelved. However, the May Day holiday is not usually on May Day itself, but on the Monday following May Day.

This festival is potentially one of the most enjoyable in the entire course of the year, but is now largely neglected. There is a great potential for its rediscovery.

Midsummer

The summer solstice has clearly been of great importance in Europe since ancient times. Alignments at Stonehenge, constructed between 3,000 and 2,000 BC, are connected to the direction of the sunrise at the summer solstice and sunset at the winter solstice. Likewise, several 5,000-year-old stone circles in Scotland were built in such a way as to create alignments between the sun and the stones at both solstices.

At least over the last 2,000 years, throughout Europe and also in north-west Africa, midsummer has been celebrated with the burning of bonfires. In many places there were also customs of throwing blazing discs through the air and rolling burning wheels down hills. In some places people made gigantic figures of wickerwork that were paraded and then burned. In other places this festival was associated with ceremonial bathing, and in England with rolling in dew, as on May Day.[39]

In Europe, the celebration of Midsummer Day was on 24 June rather than on the solstice itself, echoing the displacement of Christmas to a few days after the solstice. Since the solstice by definition is a period in which the sun rises and sets in the same place on the horizon for several days, in the absence of very precise astronomy, the definition of the solstice could only be made when the sun had risen at the same place for a few days on the north-eastern horizon before it began its southward journey; it could be recognised most clearly when it had just passed.

In many parts of northern Europe, the period of midsummer festivities extended from the eve of Midsummer Day, on 23 June, until 29 June. There were often bonfires on the nights of 23 and 28 June. The second day of celebration may have served as a kind of backup for the first in case of bad weather.[40]

The use of water as part of the midsummer rites throughout the ancient world gave sacred rivers, springs and the sea itself properties of blessing and healing. Some early Church Fathers condemned these ceremonies, but a more inclusive attitude prevailed, and the festival was itself baptised by being consecrated to St John the Baptist. Throughout Europe, Midsummer Day was known as St John's Day. The day at the end of the midsummer festival period, 29 June, was dedicated to Saints Peter and Paul.[41]

Not only in the countryside, but also in towns and cities, there were celebrations on the eve, or vigil, of St John's Day and likewise on the eve of Saints Peter and Paul's Day. Here is a description of these vigils in London from the early sixteenth century:

> There were usually made bonfires in the streets, every man bestowing wood or labour towards them: the wealthier sort also before their doors near to the said bonfire, would set out tables on the vigils, furnished with sweet bread, and good drink, and on the festival days with meats and drinks plentifully, where-unto they would invite their neighbours and passengers also to sit, and be merry with them in great familiarity, praising God for his benefits bestowed on them.[42]

In Britain, the Puritans objected to these celebrations and tried to supress them. They were more successful in towns than in rural areas, where in some places they continued into the nineteenth century. But the celebration of midsummer has undergone an astonishing rebirth in the form of summer music festivals. In England, the Glastonbury Festival, the most iconic of all these open-air celebrations, takes place at Midsummer, usually including Midsummer's Day itself.

The festival happens on Worthy Farm in the parish of Pilton, a village near Glastonbury. Pilton parish church, which dates from the eleventh century, is dedicated to St John the Baptist. Thus for many centuries the patronal festival was on 24 June. The old pattern of celebration in Pilton on Midsummer's Day has taken on a new lease of life, on a vastly increased scale.

The feast of the angels

The feast of St Michael and All Angels, or Michaelmas, on 29 September, was a quarter day in England. Michaelmas gave its name to the quarter-year period that followed it, up to Christmas. In several universities, including Oxford and Cambridge, and also in the English legal system, the term beginning around 29 September is still called the Michaelmas Term.

Michaelmas has never been a major popular festival like May Day and Midsummer, but it is an important festival nonetheless. It is the principal day in the year when the realm of the angels is celebrated. In the Roman Catholic Church there is also a special festival a few days later, on 2 October, dedicated specifically to Guardian Angels.

Christianity inherited a belief in angels from Judaism, and shares it with Islam. There are parallel beliefs in many other religions, including in India, where the *devas*, divine beings or 'shining ones', are similar to angels. Both in Hinduism and in the Abrahamic religions, angels can be good or bad; the *devas* are forces of light and *asuras* are forces of darkness, just as Michael and his angels are forces of light as opposed to Satan and his fallen angels, who are devils, forces of darkness.[43]

As the Middle East and Europe were Christianised, many pre-Christian hilltop shrines, wells and springs were dedicated to

St Michael. Examples include St Michael's Mount in Cornwall, England, and Mont St Michel in France. In fact, there are so many such places that scholars acknowledge that 'given an ancient dedication to St Michael and a site associated with a headland, hill-top, or spring, on a road or track of early origin, it is reasonable to look for a pre-Christian sanctuary.'[44]

Belief in angels is still remarkably widespread, even in modern industrial societies. In a Gallup poll in 2016, seventy-two per cent of a random sample of Americans said they believed in angels,[45] and in the same year, a survey in Britain found that thirty-two per cent believed in angels; the same proportion felt that they have a guardian angel watching over them.[46] Books on angels are often bestsellers, especially those by Lorna Byrne, an Irish woman who says she sees angels on a daily basis, and has done so since she was a child.

In 2014, at an event in Grace Cathedral, San Francisco, she said she could see many angels in the cathedral, but most were unemployed, because so few people were asking for their help.[47] I once had lunch with her in Dublin, and she told me she could see my guardian angel behind me, much larger than me.

I wrote a book with the theologian Matthew Fox about angels, called *The Physics of Angels*, in which we explored traditional ideas about angels and their relevance today. What interested me most was the old belief that there are many levels of mind or consciousness between humans and God. In the medieval, pre-mechanistic worldview, the universe was animate. The stars were alive, each with its own angelic intelligence; the heavens were full of living beings. The planets were living organisms; we still call them by the names of the old gods and goddesses, like Saturday for Saturn's Day. The earth itself was a living being, Mother Earth, an idea that has returned in a scientific form as the Gaia Hypothesis.

The angels were not humanoid figures with wings, or chubby cherubs, but the governing intelligences of nature. As the philosopher and theologian Thomas Aquinas (1225–74) expressed it, 'The entire corporeal world is governed by God through the angels. The angels are part of the universe in the sense that they do not constitute a universe on their own, but are combined with the physical creation to form one, total world . . . For the total good of the universe consists of the interrelationship of things and no part is complete and perfect in isolation from the whole.'[48]

There was no room for angels in the mechanical universe of materialism. But as we recover a sense of the living universe with mind or consciousness at many levels of organisation, including stars and galaxies (*SSP*, chapter 3), the idea of guiding intelligences at many different levels takes on a new significance.

However we think about angels, this festival provides a wonderful opportunity to acknowledge them and reconnect with them. Many Anglican and Roman Catholic churches and cathedrals celebrate Michaelmas. I usually go to Westminster Abbey, where there is a splendid evening service with priests in golden robes, incense, and amazingly beautiful singing by the choir in honour of the angels, with hymns about angelic choirs.

Days of the dead

Not everyone experiences angels or believes in them, but everyone has family members and friends who have died. In practically all traditional cultures, it is taken for granted that the living and the dead continue to interact. If living people honour the ancestors, the ancestors will help the living. Traditional Chinese and Japanese families have ancestor shrines in their houses. And all cultures revere great people who have gone before, and call on their help. Even communist states with their atheist ideologies built mausoleums to sanctify great leaders. Lenin and Stalin were entombed in Red Square in Moscow, and Mao Zedong in Tiananmen Square in Beijing. Their shrines still attract pilgrims.

In the Christian tradition in the West, the festivals of the dead are on 1 and 2 November, All Saints' Day and All Souls' Day. An older name for All Saints was All Hallows, and hence the eve of this festival, on 31 October, is called All Hallows Eve, or Hallowe'en.

In some ways, the Christian practice of praying to the saints resembles the practice of praying to gods and goddesses in polytheistic religions. Roman Catholic churches and cathedrals with their side chapels dedicated to different saints, each with their own special areas of expertise, are not unlike Hindu temples with their shrines to multiple gods and goddesses. The ancient Romans were polytheistic, too, and the great Roman temple to all the gods, the Pantheon, was consecrated by Roman priests around AD 126. It was rededicated by the Pope in 610 as a Christian church in honour of All Saints.

The festival of All Saints was set on 1 November in 835 by Pope Gregory IV to coincide with pre-Christian autumnal festivals in northern Europe. But All Saints' Day left out everyone else, and so a second festival was instituted in 988 to include the faithful departed, All Souls' Day.[49] In many Roman Catholic countries, All Saints' Day is a public holiday, and in Mexico, All Souls' Day, the Day of the Dead, is a major national festival.

The pre-Christian festival on 1 November was called *Samhain* in Ireland, marking the opening of winter. It is the direct calendrical opposite of May Day, the opening of summer. The eve of Samhain, 31 October, shared some of the properties of the eve of Beltane, when witches, fairies and elves were believed to be particularly active. In Scotland, another name for this eve was *Puca*, or goblin night; in Shetland the trolls were thought to come out from their hiding places to try and wreak havoc.[50]

As on the eve of Beltane, bonfires were burnt in many parts of Ireland, Scotland and Wales. Although it was not originally connected with the dead, it was a time when people felt the need to guard against supernatural forces. The Christian festival of the dead served to reinforce its importance, and its eve became Hallowe'en, when in medieval England, church bells were rung until midnight to comfort the souls of the departed.[51]

These festivals were severely disrupted at the Reformation. The Reformers opposed the idea of praying for the dead, as well as praying to saints for their intercession. They tried to stamp out the observance of All Souls' Day.[52] All Saints' Day remained a minor festival, but was no longer a holiday.

In many Roman Catholic countries it was, and still is, the custom for many families to visit the graves of their ancestors on All Souls' Day, and place lighted candles on their graves. In the churches, there are requiem masses, at which the dead can be remembered and prayed for by name. In 1928, the Anglican Church reinstated All Souls' Day, and such requiem masses and prayers now happen in Anglican churches, too.

In nineteenth-century Ireland and parts of Scotland and England, Hallowe'en became a mischief night on which young people played pranks, dressed up, sometimes cross-dressing. They tried to scare unpopular people with grotesque faces made by hollowing out turnips, carving faces and lighting them from within by a candle.

In America, the colonists of New England were mainly Protestant, and the Days of the Dead were more or less ignored, except for the observance of All Saints' Day in the Episcopalian Church. But Irish immigrants in the nineteenth century brought Hallowe'en practices with them. By the 1950s, Hallowe'en had become a national festivity in the United States, with children dressing up to represent ghosts, goblins or witches. Pumpkins replaced turnips, and mischief-making and house-to-house calls combined as 'trick or treat'.

A similar transformation occurred in England, initially through Irish immigration, and later because of an increasing cultural influence from America. By the end of the twentieth century, Hallowe'en had once again become a national festival.[53]

There are many ironies here. Protestant attempts to deny the ongoing relationship between the living and the dead led to a return of the repressed in the form of skeleton outfits and skull-like pumpkins. And while the festivals themselves are almost forgotten by grown-ups, the eve of these festivals has become a major time of celebration by children, with massive overdoses of sugar. Meanwhile the continued celebration of the Day of the Dead in Mexico, itself a transformation of a pre-Christian festival, has become internationally iconic.

I am thankful that so many families celebrate the eve of the festival of the dead, and that children remind us of this major festival. On All Saints' Day, I usually go to an evening service in Westminster Abbey, and I am often inspired and uplifted by the celebration of the saints, the blessed dead, with beautiful music sung by the choir, together with the communal singing of hymns in honour of the saints.

For All Souls' Day, I usually go to a requiem mass in my parish church in Hampstead, and often play the organ for this service. Every year my wife, Jill Purce, leads a week-long workshop over the Days of the Dead. Most of her participants attend this requiem. They write a list of the ancestors and departed friends they particularly want to remember and to honour. As our parish priest chants out these names during this requiem, they light candles in their memory.

This is an extraordinarily moving experience for most people who take part, including me. Similar requiems take part in Roman Catholic and Anglican churches throughout the world.

Muslim festivals and the lunar calendar

In Islam, religious festivals are completely detached from seasonal cycles. Jewish festivals are still linked to the seasons, although their meaning has been transformed. But the timing of Muslim festivals is entirely lunar, with no solar component at all.

The Muslim calendar consists of twelve lunar months, each of which begins on the new moon, which has to be seen for the month to begin. That is why the dates of the festivals in calendars are given only as probable dates, because the beginning of each month is not fixed by calculation, but by sighting. This lunar calendar means that the Islamic year is eleven days shorter than the sun-based calendar. Hence the lunar calendar cycles through the seasons: for example, the fasting month of Ramadan began on 9 December in 1999 and on 6 June in 2016.

The main Muslim feasts are *Eid al-Adha* and *Eid al-Fitr*. The first of these, the 'Feast of the Sacrifice', commemorates the willingness of Ibrahim (Abraham) to sacrifice his son. Before he almost did so, God provided a male goat to sacrifice instead. This animal sacrifice replaced a human sacrifice. If they can afford it, Muslims sacrifice one of their best domestic animals – a cow, camel, goat or sheep. The other great festival, Eid al-Fitr, occurs at the end of the fasting month of Ramadan, in which the Prophet Mohammed is believed to have received the words of the Koran.

In many Muslim countries, older seasonal festivals still persist as well. In Egypt, the Sham al-Naseem festival marks the arrival of spring in March. Many families have picnics together and musicians and dancers take over the streets. In Iran and neighbouring countries, the New Year festival of Nowruz begins at the spring equinox. This festival is ancient and long predates Islam; it lasts for thirteen days, beginning on the night of the equinox.

Benefits of holy days and festivals

There have been few scientific studies of the effects of specific festivals and holy days on health and wellbeing, but they share many features that are known to be beneficial, like having holidays from work, time to celebrate with other people, time to take part in family gatherings and feasts, time to sing, dance and give thanks, time to spend outdoors and in recreational activities, and time to attend religious ceremonies and rituals.

Participation in these holy days helps bring people together, lessens loneliness and social isolation, gives a greater sense of meaning and belonging, and through seasonal festivals links the human community to the course of the year, and to the cycles of vegetation on which all human and animal life depend.[54]

The traditions of these days themselves establish a sense of continuity over time, linking the present participants with all those who have taken part in the festival before. As I discuss in connection with rituals (*SSP*, chapter 6), by following established precedents, those who take part now are linked by a kind of resonance to those who have participated before, by a process I call morphic resonance. Patterns of activity in the present resonate with similar patterns of activity in the past, through a connection across time.[55]

The celebration of festivals activates a kind of collective memory. In the case of the major seasonal festivals, this memory includes influences from far back into the past. As I have discussed, many Jewish and Christian seasonal festivals are rooted in older pre-Jewish and pre-Christian festivals. The celebration of these festivals connects present-day participants with their ancestors and predecessors over many generations, back through history to prehistory. These deep roots are part of our humanity.

Festivals continue to evolve. The resurgence of summer music festivals taps into ancient traditions of midsummer festivities. And in India, Hindus have a remarkable ability to assimilate new features into old traditions. When I was working at the International Crops Research Institute for the Semi-Arid Tropics (ICRISAT) near Hyderabad, I arrived at my laboratory on the festival of *Durga puja* in October, and I was amazed by what I found.

On this day, people traditionally ask for the blessings of the goddess Durga on the instruments of their trade: scribes their pens, artisans their tools, bus drivers their buses, and so on. I found our laboratory was decked out with banana leaves, and the drying ovens, digital weighing scales and other apparatus were garlanded with jasmine flowers. Limes were placed under the wheels of our pick-up truck, ready to be crushed as an offering to the goddess when the truck was driven slowly forwards. As the head of the laboratory, I was asked to perform this ritual, which I did.

In the main-frame computer room, the terminals were garlanded with marigolds, incense sticks were burning, and in front of the computer was a block of stone with a coconut on it, which the American head of the Computer Services Section was asked to break as a sacrifice to ensure the blessings of the goddess on the computer for the coming year. I had never before imagined that modern technologies could be brought within the purview of ancient rituals. It was an astonishing revelation.

Holy days and festivals will continue to play an essential role in our collective life and help bind people together in families and communities, as well as connecting our human lives and stories to the more-than-human world. We now face the challenge of finding ways of adapting these ancient practices to life in modern technological societies.

Two practices of sacred time

KEEP A WEEKLY DAY OF REST

The benefits of a weekly day of rest for worship, family time, celebration, music making, playing games, reading, recreation and making love have been widely recognised throughout Jewish and Christian history, and are now supported by scientific research. But this weekly day of rest, or weekend of rest, is being undermined by secularisation, working at weekends, torrents of social-media messages and emails, and the 24/7 culture of commerce. It requires a serious effort to celebrate the day of rest, and it is very important to do so.

For people from Jewish and Christian backgrounds, weekend holidays make it possible for Jewish people to reclaim the Sabbath on Saturdays, and for people from Christian backgrounds to reclaim Sunday as a day of rest. In Muslim countries with Thursday–Friday or Friday–Saturday weekends, Muslims can easily observe Friday as a holy day.

I try to use my computer as little as possible on Sundays, avoiding emails and work-related meetings and activities. I also avoid shopping, except for immediate necessities. I usually go to church on Sunday mornings wherever I am, which gives a weekly opportunity to pray and give thanks with other people, to sing as part of a community, to be immersed in music, to participate in Holy Communion, to receive a blessing for the coming week, and to connect with a local sacred place. I try to spend Sunday lunchtimes, afternoons and evenings with my family or friends, and, especially in the summer, to be outdoors as much as possible.

If your days of rest have been eroded by work, online activities and social media, then try to reclaim them. Eliminate, or greatly reduce, screen time; switch off your smartphone and other devices. Take a technology Sabbath.[56]

The freeing of your time opens new possibilities for celebration, enjoyment, relaxation, and re-creation. Your life will have a chance to regenerate, rather than being burnt out by ceaseless work, worry and distraction. And if you have children, you can help them share in the here and now by restricting their time on screens, phones and other media. When they find that a day of rest can bring delight, and is happiest when shared with other people, they will see the point.

REDISCOVER FESTIVALS

Unless you feel strongly alienated from your own tradition, it is best to look again at the festivals in your own ancestral lineage. For Jewish people, this is relatively easy, because even secular Jews often take part in the major festivals. One of the strengths of Judaism is that it is more about shared practices than about shared beliefs. Much the same is true of Islam, and of Hinduism.

Most non-religious people in Europe and the Americas come from Christian backgrounds, and many already celebrate Christmas, and a secularised form of Easter, and these festivals are major public holidays. In my experience, the enjoyment of these festivals is greatly enhanced by participating in their religious ceremonies, rather than missing them out.

I also suggest taking part in other festivals that you may not have observed before, especially the Feast of St Michael and All Angels on 29 September, and All Saints' and All Souls' Days on 1 and 2 November. On All Souls' Day (or on the Sunday nearest to All Souls' Day), many Roman Catholic and Anglican churches hold ceremonies of commemoration in which the priest reads the names of people who have died. You can go to one of these services with a list of people whom you would like to include, such as friends or family members who have died in the previous year, or ancestors, or mentors, or anyone else you would like to honour. The people whose names you include can be commemorated in this service.

And if you can, find ways of celebrating May Day and Midsummer Day outdoors. These great festival days can include people from all religious backgrounds, and are times when the celebration of our relationship to the more-than-human world can be at its most joyful.

7
Cultivating Good Habits, Avoiding Bad Habits, and Being Kind

What use are spiritual practices? If I meditate diligently, and access states of blissful non-duality, it's good for me. I feel connected and happy. If I fast and my mind becomes exceptionally clear, I enjoy the clarity. But what good is my spiritually induced happiness to other people, or to the world?

Spiritual practices can become self-indulgent. Spirituality can be selfish, unless it helps other people and has a positive effect in the wider world.

All the spiritual practices discussed in this book, and also in my previous book, *Science and Spiritual Practices*, can be carried out within religious frameworks or outside them.

One advantage of religious frameworks is that they include traditional systems of morality. Theravada Buddhism, for example, emphasises the Noble Eightfold Path: right view, right resolve, right speech, right conduct, right livelihood, right effort, right mindfulness, and right *samadhi* or meditation. In Islam there is a strong emphasis on cultivating good character, following a moral way of living and behaving kindly.

In Hinduism, following the *dharma* brings people into harmony with nature and with their families, society and the moral order. Traditional Judaism is based on the Ten Commandments, and includes many other guides to good conduct. Christians are continually reminded of Jesus' commandment to love their neighbours as themselves. All religions have moral codes.

For people who follow spiritual practices outside a religious framework, there is no compelling reason to adopt the moral principles of any particular religion. But fortunately, there is a widespread agreement about fundamental principles among different religions and also among secular humanists. Many religions and ethical philosophies promote versions of the Golden Rule: 'Do unto others as you would have them do to you.' There is a general agreement about this fundamental principle.

In addition, there are many different culturally and religiously determined rules that form part of a given culture's morality. The very word 'morality' is rooted in the Latin *mores*, meaning customs, and some of them are specific to particular religions or cultures, like taboos on certain kinds of food, such as the prohibition on eating shellfish for Jews.

Although morality originates in customs and in religions, rather than in law, the police and law courts uphold some fundamental moral rules in secular as well as religious societies, such as prohibitions against theft and murder, and penalties for committing them.

The word 'ethical' is often used interchangeably with 'moral', but its Greek root *ethos* has a different meaning, namely 'character' or 'moral nature'. There is a distinction between rule following and the cultivation of good habits, which is where spiritual practice comes in. Following rules or customs is important in all social contexts, but is not in itself a spiritual practice. Rules provide a baseline.

For instance, to play football requires the players to obey the rules of the game. This is a necessary condition for remaining within the game. But the cultivation of skills that are needed to play good football goes beyond abiding by the rules; it requires many hours of practice. Likewise, the cultivation of virtues is a spiritual practice, because it goes beyond mere rule following. Virtues are positive skills, and require practice.

Virtues, or strengths of character, connect those who practise them to ideals, or to models of good behaviour. The more spiritual the virtues, the greater the priority they give to the good of other people and to the more-than-human world.

In this chapter, I begin with a cross-cultural survey of virtues, which shows a remarkable agreement between different religious and cultural traditions. But to what extent are virtues dependent on a religious or traditional underpinning? To look at virtues in a broad context, I explore the biological and philosophical backgrounds of social cooperation and altruism. All animal societies involve cooperation among their members, without which the societies would fall apart and their members perish. Many animals help other members of their group, especially their young, and human moralities and virtues are rooted in this evolutionary history.

Virtues as good habits

All religions promote the cultivation of virtues, as do secular humanists, and all explicitly or implicitly support the idea of character training in education, such as encouraging children to be considerate, helpful, truthful and courageous. But precisely because virtues are encouraged by religions, the study of virtues was neglected in academic psychology for most of the twentieth century.

Only since the beginning of the twenty-first century has the empirical study of virtues become established within the academic world, as part of the positive psychology movement. But in order to make this field of enquiry acceptable within a secular framework, virtues have been rebranded. They are now called 'character strengths'.

As part of a major project involving forty scholars and a review of more than 2,500 papers in scientific journals, in 2004, Christopher Peterson and Martin Seligman produced a monumental book called *Character Strengths and Virtues: A Handbook and Classification.*

This project was influenced by the *Diagnostic and Statistical Manual of Mental Disorders* (DSM), widely used by psychiatrists as a definitive reference work. As the authors write in the introduction, 'The classification of strengths presented in this book is intended to reclaim the study of character and virtue as legitimate topics of psychological inquiry and informed societal discourse.' They think of it as 'a manual of the sanities'.

A starting point for this new study of virtues was a survey of the world's religious and philosophical traditions, along with studies by anthropologists of non-literate cultures. This survey revealed remarkable agreements across the board. The six most important virtues or strengths of character esteemed all over the world are:

1. Wisdom and knowledge.
2. Courage – emotional strengths involving the exercise of will to accomplish goals in the face of opposition.
3. Humanity – interpersonal strengths that involve helping others.
4. Justice – civic strengths that underlie healthy community life.
5. Temperance – strengths protecting against excess.
6. Transcendence – strengths that forge connections to the larger universe and provide meaning.[1]

In some cultures, these virtues were not equally emphasised for everyone. In ancient Greece, the philosopher Plato summarised the four fundamental virtues as wisdom, courage, self-restraint, and justice. But like many of his contemporaries, he thought that different groups of society needed different virtues. For the ruling class, wisdom was especially important, and for the military class, courage.

In traditional Hinduism the virtues encouraged in Brahmins, the priestly caste, included wisdom and knowledge, temperance and faith, while the most desirable qualities in the warrior caste, the *Kshatriyas*, were courage, fortitude and generosity. The lower castes were assigned the more limited virtue of performing their work dutifully. And in many parts of the world, there were different sets of virtues for men and women. For example, as discussed below, in nineteenth-century Britain, kindness and compassion were esteemed in women, but in men were often seen as a sign of weakness.

In Europe in the thirteenth century, St Thomas Aquinas codified the principal virtues by combining the four cardinal virtues of Plato – temperance, courage, justice and wisdom – with the three specifically Christian virtues of faith, hope, and love. Of the cardinal virtues, Aquinas thought wisdom the most important. Above wisdom came the transcendent virtues of faith and hope. And above all was the supreme virtue of charity or love, which he described as the 'Mother and root of all virtue.'[2]

In Europe, people used to be presented with clear images of virtues and vices, with the aim of encouraging virtues and discouraging vices.

At the beginning of the fourteenth century in Padua, Italy, the artist Giotto was commissioned to decorate the walls of a chapel with a series of frescoes, each representing a different vice or virtue. On the right-hand side, he painted figures representing the classical virtues of prudence, fortitude, temperance and justice, followed by the three Christian virtues of faith, charity and hope. Directly opposite, he painted figures representing the matching vices: folly, inconstancy, anger, injustice, infidelity, envy and despair. Depictions of these vices and virtues were very widespread. Their purpose was to provide a compass by which people could steer their lives in the right direction.

By contrast, in the modern secular world there are no such depictions. As the philosopher Alain de Botton points out in his book, *Religion for Atheists*, modern secular theorists argue that the public space should be kept neutral. There should be no reminders of virtue on the walls of our public buildings or in the pages of our textbooks, because such messages would constitute infringements on our liberty. But, as de Botton points out, 'Our public spaces are not even remotely neutral. They are – as a quick glance down any high street will reveal – covered with commercial messages. Even in societies theoretically dedicated to keeping us free to make our own choices, our minds are continuously manipulated in directions we hardly consciously recognise.'[3]

These commercial messages are often incitements to greed, lust, envy, gluttony and the other traditional vices. De Botton envisages the radical possibility that in secular societies these messages could be balanced with encouragements to virtue. The state 'would try to redress the balance of messages that reach its citizens away from the merely commercial and towards a holistic conception of flourishing. True to the ambitions of Giotto's frescoes, these new messages would render vivid to us the many noble ways of behaving that we currently admire so much and blithely ignore.'[4]

This modern classification of virtues by positive psychologists is explicitly secular, and its authors describe themselves as agnostics. Nevertheless, when they first began their study of character strengths, they found that many of their academic colleagues in the United States were sceptical, pointing out that there were huge differences between cultures worldwide, and that there were many variations between religious and ethnic groups even within the United States.

The positive psychologists responded with a thought experiment. They 'tried to imagine a culture or subculture that did not stress the cultivation of courage, honesty, perseverance, hope, or kindness. Done another way, this experiment requires that we envisage parents looking at their newborn infant and being indifferent to the possibility that the child would grow up to be cowardly, dishonest, easily discouraged, pessimistic, and cruel.'[5]

The researchers also carried out a massive online international survey of more than 100,000 adults in fifty-four nations to find out which character strengths were most commonly endorsed and appreciated. In almost all cases, these turned out to be kindness, fairness, authenticity, gratitude, and open-mindedness, followed closely by prudence, modesty and self-regulation.[6]

Virtues, or strengths of character, connect those who practise them to ideals. The more spiritual the virtues, the greater the priority they give to the good of other people and to the more-than-human world. They are about connection and cooperation rather than being selfish. But which is more evolutionarily fundamental, selfishness or cooperation?

Selfishness and cooperation

In thinking about morality and altruism, there is a fundamental conflict between two points of view: top down, starting from the society as a whole, and bottom up, starting from the individual person.

Traditional views of morality generally assume that the whole society is primary; they take it for granted that human societies, like animal societies, can only work if the individuals within them are essentially cooperative. The whole society has to work together; otherwise it falls apart and cannot survive.

By contrast, in the bottom-up view, first propounded by European philosophers in the seventeenth century, most notably Thomas Hobbes, individuals are primary, and their nature is essentially self-centred.

These very different views of human societies arise from different cosmologies, or theories of the essence of nature – top down, starting from a primal unity or harmony or cosmic source, or bottom up, starting from innumerable particles of matter.

In the bottom-up, or atomistic or reductionist, worldview, the most important entities are the smallest. They are the basis of everything else. The materialist philosophers in ancient Greece, like the philosopher Democritus, assumed that the ultimate basis of reality was matter, made up of tiny indivisible particles called atoms. In the scientific revolution of the seventeenth century, several philosophers, including Thomas Hobbes, revived this atomistic theory, and built it into the foundations of modern science. It proved to be a fruitful idea, especially in chemistry.

Hobbes also applied the atomistic theory to societies. Individual people were like the atoms of societies. Hobbes popularised an ancient Roman proverb, *homo homini lupus* – 'man is wolf to man' – to deny the inherently social nature of humanity. Much subsequent European political thought has followed his view that social life does not come naturally to humans, but is an artificial construct, a voluntary arrangement between autonomous individuals who are inherently non-social. At best, they behave according to enlightened self-interest.

But ironically, Hobbes's wolf imagery ignores the fact that wolves themselves are intensely social creatures with a high degree of cooperation. Even more ironically, in the foundational myth of Rome, the twins Romulus and Remus, abandoned as babies and left to die, were saved by a she-wolf, who suckled them. Romulus later became the founder of Rome.

In this ideological context of selfish individualism, morality came to be seen as a thin veneer imposed upon the nastiness of human nature. Under Hobbes's influence, selfishness and aggression were transformed from moral vices into psychological facts.[7]

The assumption that people are essentially selfish led to a widespread cynicism about virtues in general and altruism in particular. Those who appear to be unselfish must be hypocritical, impelled by some selfish ulterior motive.[8] And as the psychoanalyst Adam Phillips and the historian Barbara Taylor put it, 'To this old suspicion, modern post-Freudian society has added two more: that kindness is a disguised form of sexuality and that kindness is a disguised form of aggression – both of which reduce kindness to a kind of covert selfishness.'[9]

The twentieth-century evolutionary theorist, George Williams, took the primal selfishness theory to its ultimate extreme when he accounted for apparently altruistic behaviour in non-human animals and in humans in terms of the selfishness of genes. From the selfish gene's point of view, it makes sense to behave selflessly towards other related individuals containing the same gene, so that more copies of the gene can survive. Altruism serves the gene's self-interest and its irrepressible desire to replicate.

Williams saw nature as inherently wretched, and explained human morality as a mere by-product of the evolutionary process: 'I account for morality as an accidental capability produced, in its boundless stupidity, by a biological process that is normally opposed to the expression of such a capability.'[10]

Within late twentieth-century biology, the selfish gene theory became the dominant fashion or paradigm. It underlay neo-Darwinian evolutionary theory, sociobiology and evolutionary psychology, and was widely popularised by Richard Dawkins. For Dawkins and his followers, selfishness is hardwired into us. Genes are intelligent, ruthless and competitive; they are like 'successful Chicago gangsters'. As Dawkins put it, these selfish genes are 'in you and me; they created us, body and mind; and their preservation is the ultimate rationale for our existence.'[11]

Similarly, within academic psychology and the social and behavioural sciences, the assumption of universal egoism reigned supreme.[12] If people felt empathy towards another's need and behaved altruistically, this was because seeing someone suffering made them feel bad, and so they tried to reduce their own bad feeling – a process called 'aversive-arousal reduction'. Or else their feelings of empathy induced them to help so that they could avoid feeling guilty for not helping.

Although it is true that some people are frequently selfish, one problem with the theory of primal selfishness is that humans have never been asocial, nor have our pre-human ancestors. We are descended from highly social apes. Living in groups was not an option for our ancestral species, but a necessity. Isolated individuals simply could not survive.

A good illustration of the fundamentally social nature of our species is that solitary confinement is one of the most extreme punishments we can think of. People usually become hopelessly depressed without social support and their health deteriorates.[13] All societies have as an ultimate sanction the expulsion of an individual who contravenes the accepted ways of behaving. This is usually fatal.

The anthropologist Colin Turnbull, who lived with a tribe of pygmies in a West African forest, described how one day there was a great commotion in the camp. A young man, Kelemoke, came rushing past hotly pursued by other youths armed with spears and knives.

Turnbull asked one of his informants what had happened. Kelemoke had been caught having sex with a girl who was his cousin, breaking an incest taboo. 'He has been driven to the forest. And he will have to live there alone. Nobody will accept him into their group after what he has done. And he will die, because he cannot live with only the forest. The forest will kill him.'

But Kelemoke was lucky. Some youths secretly took him food and 'three days later, when the hunt returned in the late afternoon, Kelemoke came wandering idly into the camp behind them, as though he too had been hunting. He looked around cautiously, but nobody said a word or even looked at him . . . He came over to the bachelors' fire and sat down. For several minutes the conversation continued as though he were not there, and I saw his face twitching. But he was too proud to speak first. Then a small child was sent over by her mother with a small bowl of food, which she put in Kelemoke's hands and gave him a shy friendly smile.

Kelemoke never flirted with his cousin again, and now, five years later, he is happily married and has two fine children.'[14]

Turnbull added, 'I never heard of anyone being completely ostracised, but the threat always is there, and that is sufficient to ensure good behaviour.'

While one strand of twentieth-century biology followed the theory of selfish genes, biologists who studied the behaviour of social animals saw things differently, emphasising the intrinsically cooperative nature of animal societies.

Countless species of animals are social. Even very simple forms of life like corals are made up of colonies of small organisms called polyps that build up large structures cooperatively. What we recognise as a coral is made up of thousands of genetically identical polyps that together form a kind of superorganism. And many species of insects are social, at least in relation to the care of their young. Earwigs, for example, often keep their young in sheltered nests or burrows, and forage for food which they bring back to feed their brood, as do predators such as paper wasps.[15]

They look after their immediate family. But some insect species take social organisation to levels unparalleled by any other kinds of animals, except humans. They illustrate just how far social cooperation can go, even though almost all the members of the society are sterile and leave no personal descendants.

Selfless cooperation within insect societies

Some of the most remarkable of all non-human societies are those of termites, ants, wasps and bees, which contain thousands, even millions of individual insects. Among the most complex are colonies of leaf-cutting ants in the genus *Atta*, which live in Central and South America. Some *Atta* colonies contain up to eight million individuals, with a broad array of castes and sub-castes and a great range of sizes. The smallest minor workers weigh only 1/200th of the major workers.

These ants collect huge quantities of leaf material that they cut from plants, then chop up into small fragments and use for cultivating fungal gardens in their nests. The fungi digest the leaves, which the ants themselves cannot do. Large foragers cut pieces of leaves and bring them back into the nest, where smaller workers clip them into fragments.

Within minutes, still smaller ants take over and crush these fragments into moist pellets to which they add faecal droplets and insert them into a mass of similar material. Next, even smaller workers pluck loose strands of fungus from places of dense growth and plant them on the newly constructed surfaces. Finally, the smallest and most abundant workers patrol the beds of fungal strands, delicately probing them, licking their surfaces and plucking out spores and threads of alien species of fungus.

In addition to this complex division of labour, individual ants perform a succession of tasks as they grow older. The younger workers are usually inside the nest and the older ones work outside. The very smallest ants look after the fungus inside the nest when they are young, and as they grow older start working at the harvesting sites, where they defend the leaf carriers from attacks by parasitic flies, which attempt to lay eggs on the ants' bodies. Many of them do not walk back to the nest on their own, but hitchhike on the leaf fragments being carried to the nests.

If an *Atta* colony is attacked, most types of older worker take part in defence, but the response depends on the kind of attack. If a large enemy such as a vertebrate predator threatens the colony, gigantic soldier ants with sharp mouthparts spring into action, but if other ants attack the colony, the smaller workers respond.[16]

Just as the millions of cells within animal bodies are coordinated to work harmoniously together for the good of the whole, so these individual insects within complex societies behave like cells within a superorganism, and their activities are coordinated in a way that goes beyond the mental capacities of any individual within the group. They have very small nervous systems compared to vertebrates, yet their social complexity is vastly greater than that of many species with far larger brains.

Even if we think of each individual as following an inherited set of habits, this would not explain how their activities are coordinated so effectively with each other, so that they can respond to attacks and to new circumstances. Thinking of them as superorganisms emphasises that the whole is greater than the sum of the parts.

The workers are sexually sterile, so they are not doing all this work for their own personal reproductive success, but rather for the survival and reproduction of the colony as a whole. Some insects directly help other members of their group by rescuing them, as in several species of sand-dwelling ants. In laboratory experiments, the researchers trapped some workers in nylon snares to see what their nest mates would do.

At first, they tried to rescue them by digging in the sand or pulling their limbs through the nylon mesh, but when this did not work, they began to bite precisely at the nylon, cutting through it so they could release their trapped fellow workers. The rescuers discriminated between victims; they helped only distressed nest mates and not motionless ants, or members of different colonies, or ants from different species. They displayed what the researchers described as 'sophisticated helping responses towards familiar distressed individuals'.[17]

This behaviour was certainly altruistic, in the sense that it helped others while giving no direct or immediate advantage to the rescuers. Among social animals, even those with very small brains, individual behaviour is often coordinated in such a way that the group as a whole has a better chance of surviving and reproducing.

An insect colony is not a voluntary association of free egotistic individuals, where individual selfishness comes first. The individual insects are totally dependent on the colony as a whole for their existence and survival, and the field or system of the colony as a whole organises their activity. Their individual lives are subordinated to the needs of the whole, and they follow a kind of unconscious, instinctive morality and sense of duty. They are like cells in the body of an organism and, like cells, are very similar genetically.

Similar coordinating principles are likely to be at work to some degree in all animal societies, including human societies, even though the individuals are less closely related than members of an insect colony. Our experiences of social pressure, and even of conscience – that inner sense of right or wrong – seem to be derived from the fields or organising principles of the social groups to which we belong, as discussed below.

The evolutionary roots of human morality

The fact that social groups are inherently cooperative enables us to recognise continuities between societies of non-human animals and societies of humans. Morality is not imposed by human culture on inherently isolated and selfish individuals, but emerges from the very nature of social organisation.

Charles Darwin, who was not a neo-Darwinian, made this point explicit: 'Any animal whatever, endowed with well-marked social instincts, the parental and filial affections being here included, would inevitably acquire a moral sense or conscience, as soon as its intellectual powers had become as well developed, or nearly as well developed, as in man.'[18]

First and foremost, Darwin thought, 'the social instincts lead an animal to take pleasure in the society of its fellows, to feel a certain amount of sympathy for them, and to perform various services for them.'[19]

In animal societies, social behaviour depends on a range of emotions, including anger and retributive emotions, as well as more positive emotions. Helpful behaviour has been observed in many kinds of animals, and is often immediate and seemingly unreflective. The instinct to help others probably first arose in the context of parental care, where parents feed and protect their vulnerable offspring, but in many social animals these responses stretch far beyond the realm of parental care to relations between unrelated adults. Many animals help other members of their group, rescuing those in distress and defending others from attack. Some even rescue members of other species.

In the scientific literature, when one individual's actions benefit another without an immediate benefit to itself, it is called prosocial behaviour. Such behaviour is often associated with the release of the so-called love hormone, oxytocin, which seems to promote helping behaviour. In an experimental study with pinyon jays, a highly social species, individuals were paired with another jay in an adjacent cage, and given food, while their companion was not. They spontaneously shared some of this food with the bird that had not been fed, but they shared even more food when they were dosed with mesotocin, the equivalent of oxytocin for birds.[20]

Sometimes helping behaviour occurs not only in response to immediate needs, but also over sustained periods. In a study of chimpanzees in the wild, over a two-day period an adolescent male helped an injured mother to whom he was not related. The mother was unable to keep up with the rest of the group and frequently stopped, leaving her baby on the ground. The young male picked up the infant and carried her as the group moved. He did not carry the baby again after this journey was completed, and he had not been seen to carry it or any other baby before this incident. Clearly this young male recognised the mother's state and her needs.[21]

The biologist Frans de Waal, who has spent many years studying the behaviour of chimpanzees and other primates, points out that the evolutionary origin of the capacity for sympathy is no mystery:

All species that rely on cooperation – from elephants to wolves and people – share group loyalty and helping tendencies. These tendencies evolved in the context of a close-knit social life in which they benefited relatives and companions able to repay the favour. The impulse to help was therefore never totally without survival value to the ones showing the impulse. But, as so often, the impulse became divorced from the consequences that shaped its evolution. This permitted its expression even when payoffs were unlikely, such as when strangers were beneficiaries. This brings animal altruism much closer to that of humans than usually thought.[22]

Some animals even show behaviour that we would identify as forgiveness. Chimpanzees often kiss and embrace after fights, and these reconciliations help preserve peace within the community.[23] Chimpanzees also show empathy towards members of their group that are suffering, for example by consoling the victims of fights.

Those offering comfort do not usually seem distressed themselves before they approach the victim, suggesting that the main role of this behaviour is not for the benefit of the comforter, but for the alleviation of the other's distress. Similar consoling behaviour towards victims of aggression also occurs in wolves, elephants, dolphins, rats and birds of the crow family, including ravens, rooks and jackdaws.[24]

As we saw in Chapter Two, many dogs, cats and other pets comfort their owners when they are in need, responding empathetically to their distress. And on the internet, videos showing empathetic behaviour by animals to members of other species are immensely popular, including footage of a hippopotamus rescuing a baby zebra from drowning in a river,[25] and an orang-utan saving a drowning bird.[26]

Not all behaviour within social species is cooperative: some is competitive, especially in relation to food supplies and sexual partners. But cooperation is predominant; otherwise the social group would not hold together. And cooperative behaviour is primarily restricted to members of the same social group. Ants rescue ants from their own colony, but not from other colonies; parents primarily protect their own young, rather than the young of others; and mutual help generally depends on reciprocal relationships within the group. Acts of compassion to strangers from the same species or from distant species are rare exceptions, which is why they attract so much interest.

Morality and conscience

The same principles apply to human societies. Cooperation and mutual help are strongest within family and tribal groups, and there is often rivalry or even warfare between different groups, usually in competition for resources or territory. Different societies have different moralities or customs, but all depend on fundamental principles of helping and not harming others within the social group.

The German family therapist, Bert Hellinger, has pointed out particularly clearly how these fundamental features of our social relationships are experienced through feelings of guilt and innocence: 'With every action we take that affects others, we feel guilty or innocent. Just as the eye discriminates continually between light and dark, so too an inner organ continually discriminates between what serves and what hinders our relationships.'[27]

In our social relationships, Hellinger summarises our most basic needs as:

- The need to belong, or the need for bonding.
- The need to maintain a balance of giving and taking, to maintain an equilibrium.
- The need for the safety of social convention and predictability, in other words for order.

These needs both constrain our relationships and make them possible. And because they are so important for our survival, we continually monitor what serves and what hinders our relationships. When our actions endanger our relationships, we typically feel guilt, and when our actions serve our relationships we feel freedom from guilt, or innocence. Hellinger points out that these feelings of guilt and innocence are primarily concerned with maintaining social cohesiveness, rather than being concerned with higher moral values. And since we belong to several different social groups, including families, institutions, teams, work colleagues and local communities, we experience different moralities and different kinds of guilt and innocence within each group.

We experience guilt and innocence differently according to the needs being served:

1. When our belonging is endangered, guilt feels like exclusion and alienation. When our belonging is well served, we feel innocence as intimacy and closeness.
2. When our giving and taking are not balanced, guilt feels like indebtedness and obligation. When they are well served, we feel innocence as freedom.
3. When we deviate from a social order, guilt feels like transgression, with a fear of consequences or punishment. When we are acting in accordances with the social order, we feel our innocence as conscientiousness and loyalty.

Because we belong to different social groups, what makes us feel innocent in one may make us feel guilty in another. In a gang of thieves, members must steal, and they do so with a clear conscience. In other contexts, they may feel guilty about their behaviour. And although acting in the service of belonging binds us to one another in a group, it can also lead us to exclude those who are different and to deny them the right to membership.

In racial, religious and national conflicts, members of a group can commit atrocities with a clear conscience against others who belong to different groups. As Hellinger points out,

> Guilt and innocence are not the same as good and evil. We do destructive and evil things with a clear conscience when they serve the groups that are necessary for our survival, and we take constructive actions with a guilty conscience when these acts jeopardise our membership in these same groups.[28]

A striking example of these principles was provided by the Truth and Reconciliation Commission set up by Nelson Mandela's government in South Africa. The government granted an amnesty to members of the former secret police who were willing to testify publicly about their former activities. Under the apartheid government, they had tortured and murdered people in the belief that they were doing good, acting in justified defence of their endangered nation. But after the change of government, and having been granted amnesty, in the new political context many viewed their former activities differently, and showed genuine and deep remorse.[29] As the social context changed, so did their feelings of guilt and innocence.

In traditional societies, moralities are largely customary. They are not codified as written rules and laws. In religions, moral codes, like the Ten Commandments, make explicit the moral rules that members of the religion are supposed to follow. In civilised societies, laws enforce some traditional aspects of morality, like prohibitions against stealing and murder, but law and morality do not overlap completely. Some actions may be immoral but not illegal, like neglecting family members who are in need, or being hurtfully rude to vulnerable people; others may be illegal but not immoral in some people's eyes, like taking psychedelic drugs. And the moralities of societies and religions change and evolve.

In the year 1800, British slave traders were making fortunes, and slavery was legal and widely practised in the Caribbean and in the southern parts of the United States, as well as in Central and South America. Slavery was supported by many Christians, and also by many Enlightenment intellectuals, including Thomas Jefferson, himself a slave owner. By the year 1900, slavery was illegal and widely regarded as immoral by Christians and non-Christians alike.

However, obeying laws and behaving morally in accordance with the prevailing standards of the time are not in themselves spiritual practices. They are necessary to avoid punishment, exclusion or social disapproval. Virtues and vices are different. Virtues are to do with the character of individuals. Virtues are good habits, and they can be cultivated. Vices are bad habits, and they can be avoided.

Vices, like drug addiction and predatory sexual behaviour, can easily lead to conflicts with the prevailing system of morality and the rule of law, but so can virtues. The philosopher Socrates was condemned to death in Athens and poisoned with hemlock; Jesus was crucified; Mahatma Gandhi and Martin Luther King were assassinated. Many martyrs have died for their faith in principles that transcend the prevailing laws and morals of their time.

Vices as bad habits

Just as there are classifications of character strengths, there are taxonomies of character weaknesses, or vices, or bad habits, or sins.

The word sin is now unfashionable, because of its religious connotations. But it is helpful precisely because it brings out a spiritual dimension. From a spiritual perspective, sins are not solely about breaking moral rules. They separate us from each other and also separate us from God, or ultimate reality. They are essentially divisive, whereas the virtues are connective.

Each religion and culture has its own understanding of vices, and there is often much overlap between traditions. For instance, in Islam, one classification of vices links them to deficiencies or excesses of qualities that are good in moderation.[30] The eight vices of deficiency and excess are:

Stupidity, a deficiency of wisdom.
Slyness, an excessive use of the intellect when it is inappropriate.
Cowardice, a deficiency of courage.
Foolhardiness, an excess of courage, or recklessness.
Lethargy, a deficiency in using things that the body needs.
Rapaciousness, an excess in sexual activity, eating, drinking and other sensual pleasures.
Submissiveness, a deficiency of justice, accepting oppression and tyranny.
Tyranny, the opposite extreme of submissiveness, namely the oppression of oneself or others.

In the traditional Christian culture of Western Europe, the standard taxonomy recognised seven deadly or cardinal or capital sins: pride, envy, greed, anger, gluttony, lust and laziness. Some were more mental or spiritual, like pride, envy, and avarice; others were more bodily, like gluttony, lust and sloth. Anger could be either – hot and bodily, as in rage, or cold and mental, as in bitterness and resentment.

PRIDE

The spiritual sins were worse than bodily or carnal sins, and pride or arrogance the worst of all. Pride was often called 'the root of all evil'. It was the primary sin of Satan, the angel whose pride inflated his own significance and made him attempt to rival God. According to Thomas Aquinas, the devil's desire to be godlike was not in itself a sin; the problem was that he wanted to do it by the force of his own nature, rather than through God's grace.[31]

When we meet people who are arrogant, we can feel how they separate themselves from us and from others and see how they exaggerate their own importance; they cause conflict and put themselves at odds with their surroundings.

Our collective arrogance, with the godlike powers conferred by science and technology, puts us all into conflict with our environment. Our pride has planetary effects, and separates us from the ecologies on which we depend.

ENVY

Pride is the worst of the spiritual sins, but envy is not far behind. Through envy, other people's good fortune makes us unhappy. Envy corrodes our happiness, separates us from the happiness of others and breeds malice towards them. As Matthew Fox puts it, 'Envy is damaged relationships; it sees the other as a problem, as an obstacle, as competition. Instead of cooperation flowing from community, envy gives rise to estrangement and antagonism, rivalry and contention, strife and opposition.'[32]

ANGER

The cardinal sins reinforce each other. Pride or self-centeredness can lead to vanity, and vanity to envy of others' good fortune, and envy to anger and resentment.[33] Anger is a normal emotion and sometimes healthy, but when prolonged into resentment and desire for revenge, it is deeply destructive of our own peace and of our relationships with others. It gives rise to bitterness, passive-aggressive behaviour, violence, destruction, including self-destruction, and insecurity.[34]

GREED

Greed or avarice is a spiritual sin, as opposed to gluttony, which is about bodily desires. We all have basic needs and there is nothing wrong with fulfilling them, but allowing a greed for more and more money, or power, or possessions, to dominate our lives is separative; it cuts us off from other people and from a connection with a greater sense of wholeness. As Aquinas pointed out, 'the greed for gain knows no limit and tends to infinity.'

The desire for infinity is a spiritual desire, part of a desire for God or spirit. Greed looks for the infinite in the wrong places, because objects and money can never satisfy it.[35] And it leads to violence, restlessness and injustice, selfishly taking what should be shared with other people.

GLUTTONY

Gluttony is about bodily desires, primarily eating and drinking. Unlike avarice, its goals are not infinite, because there is a limit to how much we can eat or drink. The problem is not the basic need for food or drink, or the enjoyment of eating and drinking, but obsessive excess.

Collectively, we have created a society in which we are defined as individuals by our consumption; we are consumers. And we are literal as well as metaphorical consumers. More than a third of the world's population is overweight or obese. But gluttony is not only about compulsive eating, or the compulsive drinking of alcohol, but also about many other kinds of addiction, including addiction to drugs, gambling, television, social media, or shopping. Whether we call addictive appetites sins, vices, or character weaknesses, their destructive influence is obvious to others, and eventually even to the person who is an addict. Gluttony is not only socially and environmentally destructive, but also self-destructive.

SLOTH

Sloth or laziness can take an obvious physical form, and is often linked to gluttony. Someone who eats too much, drinks too much, takes no exercise and sits around watching hours of television may represent sloth in its most obvious bodily form, but it takes several other forms. Its spiritual form is traditionally called *acedia*, or spiritual sloth, laziness about the deeper sense of connection. Thomas Aquinas defined acedia as 'the lack of energy to begin new things.'

As Matthew Fox summarises it, acedia is 'a kind of ennui, depression, cynicism, sadness, boredom, listlessness, couch-potato-itis, being passive, apathy, psychic exhaustion, having no energy.'[36] Acedia is lonely. It is about being cut off from other people, and from our relationship to the more-than-human world. The opposite of acedia is joy.

LUST

Sexual desire is part of our nature, an important part of healthy human and animal life, and necessary for the survival of our families and our species. None of us would be here without it. Lust is not bad in itself. The sexual revolution in the 1960s rejected traditional sexual restrictions and moral codes, and was a kind of liberation. But the negative aspects of lust have not gone away. Lust is still a vice, or a sin, or a weakness of character, when it becomes obsessive and selfish, as in sex addiction, rape, paedophilia, sexual harassment and addiction to pornography.

Addictions

The seven deadly sins were classified as deadly, cardinal or capital because they were, and are, addictive, and cause increasing separation between people, and between people and the more-than-human world.

There are many other forms of selfish behaviour and social vice, including compulsive lying and excessive fearfulness, but it is not my purpose here to make a comprehensive list of defects of character. Their relevance to spiritual practices is in principle very simple. For our own sake, and for the sake of others, it is best to avoid bad habits. If we are trapped or enslaved by them, then we should try to give them up. We should break free.

Of course, this is much easier said than done. There are many therapists devoted to helping people on this liberating path, as well as traditional religious practices.

The traditional religious approach is to acknowledge these faults, confess them, and ask for forgiveness and healing. Countless people have been helped by what they experience as God's love, grace and forgiveness.

Secular therapies still include a form of confession – to psychotherapists rather than priests – and attempt to create a larger framework of understanding through techniques like cognitive behavioural therapy. But purely secular methods are often ineffective. Some of the most effective methods explicitly recognise that addictions have a spiritual dimension. Many people find a path to liberation though twelve-step programmes, originally developed in the context of alcohol addiction, through the organisation Alcoholics Anonymous.

There are now more than 130 twelve-step programmes, including Gambling Anonymous, Narcotics Anonymous and Sex Addicts Anonymous.[37] All share common steps. At the beginning of the process, there is an acknowledgement that solving the problem requires a power greater than ourselves, whether it is called God or not. These programmes also recognise that we depend on our relationships to other people; we cannot do it on our own.

One of the most eloquent popularisers of the twelve-step approach is the comedian Russell Brand. In his book, *Recovery: Freedom from our Addictions*, one of his points is that there is not a special category of people who are addicts, like alcoholics and heroin addicts, while most other people are addiction-free. The question is, where are we on the addiction spectrum? Are we social-media addicts? Nicotine addicts? Shopping addicts? Gambling addicts? TV addicts? Addicted to obsessive relationships?

The commercial influences on modern society encourage a wide spectrum of addictive behaviours that cut us off from each other and from the spiritual world. They separate us. But ironically, they arise from a desire for connection. As Brand puts it, 'When I focus on the object of my addiction in any given moment, it is because I believe it will give me relief from disconnection. Even if it will ultimately make things worse, I will feel the connection. This is why addicts relapse even though they have strong evidence that the action will not be successful.'[38]

As I say, twelve-step programmes depend on re-establishing a spiritual connection with God, or ultimate reality, or a Higher Power. Step 1 is a recognition that we are powerless over our addictions. Without this, we stay in our addictions. Anyone who takes step 1 can then move to step 2. As described by those who have been liberated by this programme: 'We believed that a Power greater than ourselves could restore us to sanity.'

And then to step 3: 'We made a decision to turn our will and our lives over to the care of God *as we understand Him.*' The re-establishing of a connection to God, or ultimate reality, or The All, is the key to freedom. By connecting with the ultimate source, the repetitive need to seek temporary connections through destructive addictions is removed.

The last step is to help others: 'Having had a spiritual awakening as a result of these steps, we tried to carry this message to addicts, and to practise these principles in all our affairs.'[39] In Brand's paraphrase: 'Look at life less selfishly, be nice to everyone, help people if you can.'[40]

Cultivating virtues

The four cardinal virtues – courage, temperance, justice and wisdom – are assets in all kinds of human society. They are fundamentally related to living in a community.

Courage is of positive value in non-human societies, as well as human societies, even if it is entirely instinctive and habitual, as in soldier ants attacking enemies in defence of the colony, often at the risk of their own lives. Many mothers throughout the animal kingdom are prepared to die in defence of their young, and in some species so are fathers.

Similarly, temperance or moderation is a virtue everywhere. It is the middle path between undesirable extremes. And moderation applies to the other virtues, too. Courage, Aristotle said, is a virtue because it is the middle way between rashness on the one hand and cowardice on the other.

Justice or fairness is also implicitly esteemed in many animal societies. Social life depends on reciprocal exchanges. In human societies, receiving and giving gifts are an obvious example of reciprocal exchange. As Hellinger pointed out, a sense of fairness and reciprocity is deeply embedded in our own social nature. In developed societies, vastly complex and powerful legal systems are based on this need for fairness.

Wisdom – acting appropriately in accordance with experience – is of survival value for any society, of whatever kind. Acting appropriately works better than acting inappropriately, or acting on the basis of dysfunctional habits or random impulses. The four cardinal virtues are fundamental to many kinds of social groups, both human and non-human.

The traditional virtues of faith, hope and charity are explicitly spiritual or transcendent and therefore not applicable to everyone in the same way as the cardinal virtues. In the cross-cultural list that summarises the six main categories of virtue (p. 211), faith and hope fall into the category of transcendent virtues, which forge connections to the more-than-human world and provide meaning. Charity or love comes under the heading of 'humanity', or helping others.

In the Judeo-Christian tradition, the essential feature of faith is to dare to go into the unknown in a spirit of trust. It does not mean blind belief in dogmas. In the New Testament, in the Epistle to the Hebrews, faith is described in the light of Jewish history and practice:

> Now faith is the assurance of things hoped for, the conviction of things not seen . . . By faith Noah, warned by God about events as yet unseen, respected the warning and built an ark to save his household . . . By faith Abraham obeyed when he was called to set out for a place that he was to receive as an inheritance; and he set out, not knowing where he was going. (Hebrews 11: 1–2; 7–8)[41]

One way that God works in history is through human faith. This is a central Judeo-Christian message, and in a secularised form is now part of the ideology of progress and economic development in all modern societies. By contrast, almost all traditional societies and ancient civilisations believed that nature, history, and human lives moved in cycles. There was no overall progression. Empires rose up and fell down. And many believed that life itself was cyclical, with endless rounds of reincarnation.

Instead, Jews and Christians believed that God worked through history, ultimately leading people of faith to something better – to a collective salvation. He interacted with people through their faith, and their faith could change the course of history.

For many centuries, exiled Jews prayed at Passover, 'Next year in Jerusalem'. They had faith that they would one day return to the holy city. Then, sure enough, with the establishment of the state of Israel in 1948, this came true, at least in a political sense, with complex consequences.

Without this faith, the Jews would have been like many other conquered and exiled peoples who lost their land and their identity, and ultimately merged in with the dominant cultures. Their faith in God's action through the events of history is what made Jewish people so different from other ancient cultures. Atheists discount any possible role for God in what happens, and ascribe it to self-fulfilling prophecies. But self-fulfilling prophecies often work, and faith is often effective.

The modern cult of progress is a secular derivative of the Judeo-Christian faith in the coming of a new order of time, a transformation of history. In medieval Europe, in a succession of millenarian movements, prophets proclaimed the coming of a new age in a cataclysmic upheaval.[42] These movements pictured a collective salvation, enjoyed not by individuals but by the community of the faithful, and saw the coming transformation not merely as an improvement, but as a total transformation of life on earth.

Religious millenarian movements believed that this transformation would be brought about by miraculous divine interventions, but in secular revolutionary movements, like the French Revolution and the Bolshevik Revolution, humankind itself was to be the agent of this change.[43] One expression of this belief in the transformation of humanity and of the earth was built into the foundations of the United States. The motto on the Great Seal of the United States, shown on every dollar bill, is *Novus ordo seclorum*, 'new order of the ages'.

Thus the Judeo-Christian vision of the transformation of humanity and the earth morphed into a secular rather than an explicitly religious faith. Instead of God's direct intervention in history, or the rule of a Messiah, or the second coming of Christ, salvation would come through science, technology, reason and economic growth.

The principal prophet of this vision was Francis Bacon, writing in the early seventeenth century, when mechanistic science was still in its infancy. His aim was nothing less than 'to endeavour to establish the power and dominion of the human race itself over the universe.'[44] He envisaged this process as being led by a kind of scientific priesthood. The philosophers of the Enlightenment in the late eighteenth century shared this faith in science, reason and progress, and took it further.

Nineteenth- and twentieth-century communists believed in it. The ideologists of neo-liberal capitalism still believe in it. And faith in progress works; it is self-fulfilling. It is visibly transforming the world around us. And if we do not like the way things are going, either we do nothing, or we have faith that a better world is possible. If we want to bring about positive changes, we need to have faith in something we are working towards.

The traditional Christian faith was in the coming of the kingdom of heaven, in which human history would be transformed. This is still what Christians pray for every time they say the Lord's Prayer: 'Thy kingdom come'. A spiritual faith underlies the vision of secular progress, and still complements it.

The virtue of hope is more fundamental than faith. It seems to be built into life itself. Millions of sperm or pollen grains attempt to fertilise an egg, even though the vast majority will fail. Every fertilised egg attempts to become a new organism, even though most will die. Countless seedlings sprout in the spring. Many perish young. But all the sprouting seeds seem hopeful. And among humans, too, optimism is a virtue. It literally means hoping for the best, from the Latin *optimum* = best thing. And when coupled with the transcendent virtue of faith, hope can be history changing.

At first sight, love or charity or kindness, the third of the spiritual virtues, seems the least transcendent, the most basic of all the virtues. Cooperation is fundamental in all animal societies. Ants and other social insects work together in huge colonies for the good of the whole. The workers have no ulterior personal motive and they leave no offspring. Soldiers are prepared to die for the sake of their immediate kin. When honeybees sting a potential enemy, they die: their guts are attached to their barbed stings. They are suicide stingers. And the queen at the centre of an insect colony is not an arbitrary tyrant revelling in her power. Her role is to lay eggs, and keep laying eggs, thousands or millions of them.

The members of the colony are mutually dependent. They behave as if they continuously put the greater good above their own personal pleasure and convenience. They behave instinctively. Perhaps they even experience love as an emotion. In any case, they act unselfishly, and like cells in a body, they are parts of a single organism.

Throughout the animal kingdom, in some species parents leave their young to fend for themselves, like the tiny caterpillars that hatch out from the eggs that a female peacock butterfly lays on a nettle leaf. But in many species, parents help and nurture their young; they feed and protect them. We do not know about their subjective feelings, but they behave as if they love their young. Some mothers and fathers will, if necessary, fight to the death to protect their offspring from attack. Likewise, in human societies all over the world, many parents devote much of their time and energy to providing for their young and protecting them.

In human social groups, there has been, and still is, a willingness to sacrifice one's life for the sake of the group. Anthropologists have studied a wide range of societies to find out what factors led to extreme self-sacrifice. These surveys included tribes in Papua New Guinea, Libyan insurgents, Muslim fundamentalists in Indonesia, Brazilian football hooligans, and historical examples like Spartans at the battle of Thermopylae and kamikaze Japanese pilots in the Second World War.

The common factor was that 'extreme self-sacrifice is motivated by "identity fusion", a visceral sense of oneness with the group arising from intense collective experiences (e.g. painful rituals or the horrors of frontline combat) or perceptions of shared biology. In ancient societies, fusion would have enabled warlike bands to stand united despite strong temptations to scatter and flee.'[45]

Within an animal or a person's immediate social group, mutual help is normal and instinctive, generally speaking. Without it, the group would not survive. And, sure enough, experiments by positive psychologists support the view that 'true altruism – acting with the goal of benefiting another – does exist and is a part of human nature.'[46]

Kindness, like most other character strengths, is an instinct before it is a virtue. It comes naturally when it is directed towards one's own kind, towards kin. Throughout the animal kingdom, social animals are generally kind to their offspring, and often to members of the wider social group. They are kind because they are kin.

Kindness

As we have seen, cooperation and helping others are built in to all social animals. They are natural ways of behaving in countless animal species. They are part of our inherited nature. They spring up spontaneously in young children. And they are encouraged by all religions, and also by secular humanists.

The Christian vision of love as a spiritual virtue builds on the basic biological foundations of kindness to kin. The primary models of divine love are parental. God is a loving father and the Blessed Virgin Mary is a loving mother. Jesus, the prototype of Christian kindness, was the son of both. He modelled a kindness that was extended to strangers and to social outcasts.

In the Koran, Allah himself is repeatedly described as 'kind and merciful'[47] and he encourages people to be kind. In the collection of sayings of the Prophet, the Hadith, there are many expressions that confirm its importance, for example, 'Allah is kind and He loves kindness in all matters.'[48] In Tibetan Buddhism, the ideal of kindness is taken to embrace not only all humans, but also all sentient beings.

Yet in some modern societies, these feelings are inhibited as people grow up. In nineteenth-century Britain, egotism came to eclipse kindness as a social norm, at least for men. In a world of competitive individualism, kindness was a sign of weakness. Egoism was the engine of human progress. As Thomas Malthus put it in his *Essay on the Principle of Population* (1798):

> It is to the apparently narrow principle of self-love that we are indebted for all the noblest examples of human genius . . . for everything indeed that distinguishes the civilised from the savage state; and no sufficient change has as yet taken place in the nature of civilised man that enables us to say that he . . . may safely throw down the ladder by which he has risen to this eminence.[49]

In other words, civilised man had better go on being selfish, or else his own elevated position would be threatened. Malthus thought that philanthropy designed to help the poor would do more harm than good. They would breed until overpopulation led to universal want. Nearly a century later, the atheist philosopher Nietzsche was even more dismissive of benevolence and philanthropy. In his critique of the roots of morality, in *The Genealogy of Morals* (1887), he regarded 'the inexorable progress of the morality of compassion which affected even the philosophers with its illness, as the most sinister symptom of the sinister development of our European culture.'[50]

In nineteenth-century Europe, kindness was steadily down-graded from a universal ideal to the specialised realms of social subgroups, like romantic poets, clergymen, charity workers and, above all, women.[51] Kindness was feminised. And it remains so to this day. But it is inherent in all our natures, whether masculine or feminine, even though it may have been partially suppressed by our education and our enculturation into the individualistic and competitive modern world.

The research of positive psychologists is once again proving the obvious. In a series of experimental tests, some of these psychologists looked at the effects of committing 'random acts of kindness' on the happiness of those who did them, compared with a control group who made no effort to be more kind than usual. Sure enough, being kind made people happier.

But, interestingly, in a study in which participants were asked to perform five acts of kindness per week, when they did them all on the same day, say on Monday, they were significantly happier than when they spread them out over the week. Perhaps the effects were diluted when they spread them out, and less noticeable against the background of regular acts of kindness.[52]

Another study, carried out in Japan, found a close link between kindness and happiness in everyday life. Kind people were happier and had happier memories. Simply by remembering and counting up acts of kindness, people became happier and more grateful.[53] Other benefits of being kind to others included the decrease of pain by stimulating the release of endorphins – natural painkillers – within the brain; reducing stress, as measured by lower levels of cortisol, a stress hormone, in the blood; reducing anxiety and depression; and lowering blood pressure.[54]

The fact that people who are kind tend to be happier might fuel the suspicion that they are kind *because* they want to be happier, have lower blood pressure, and suffer less stress and anxiety. By being kind, they may in fact be acting selfishly, in accordance with enlightened self-interest. But very young children are often spontaneously cooperative, implying that this sort of behaviour is deeply embedded in human nature.[55] The psychologist Michael Tomasello and his colleagues found that infants as young as fourteen months spontaneously helped other people by handing over objects that they were reaching for unsuccessfully, and worked together towards shared goals.[56]

No modern research would be complete without brain scans, and sure enough, when people behaving cooperatively had their brains scanned, there was a consistent activation of areas of the brain associated with rewards, including the caudate nucleus, the orbitofrontal cortex and the anterior cingulate cortex.[57]

If, as a spiritual practice, we want to widen the range of our kindness, then we have to widen the range of those to whom we feel related. And in so far as we feel connected to God, the ground of all being, the supreme reality, the All, we feel a great expansion of brotherhood and sisterhood with other people, and also with other animals, plants, places, the earth and the heavens.

Compassion, kindness, and benevolence to all people, however unrelated, are a feature of all the universalist faiths, Buddhism, Christianity and Islam, and of secular humanism as well. As the Dalai Lama put it:

> Genuine compassion is based on the recognition that others have the right to happiness just like yourself, and therefore even your enemy is a human being with the same wish for happiness as you, and the same right to happiness as you. A sense of concern developed on this basis is what we call compassion; it extends to everyone, irrespective of whether the person's attitude toward you is hostile or friendly.[58]

To make a practice of being kind, to have time and energy to be compassionate and loving beyond our immediate social circle, is a spiritual practice that improves with practice. It has to be rooted in a sense of connection. The Dalai Lama explicitly recognised the affinity of the Christian and the Buddhist versions of the Golden Rule:

> For me, the main message of the Gospels is love for our fellow human beings, and the reason we should develop this is because we love God. I understand this in the sense of having intimate love. Such religious teachings are very powerful to increase and extend our good qualities. The Buddhist approach presents a very clear method. First, we try to consider all sentient beings as equal. Then we consider that the lives of all beings are just as precious as our own, and through this we develop and sense of concern for others.[59]

This Buddhist emphasis on the equality of all sentient beings sets the highest possible standard. Christian, Muslims and secular humanists all believe, in principle, in the fundamental value of human lives, and the importance of kindness to others, whatever their social status, race or religion. Some go much further than others, especially those who care for people who are downtrodden, impaired, sick, imprisoned, orphaned, widowed, aged, mentally ill and emotionally disturbed.

But even some of the most saintly people do not extend their compassion and kindness to all sentient beings, including insects. Tibetan Buddhists are often meat eaters. Yet many people care deeply about the wellbeing of some non-human animals, especially their own dogs, cats, parrots or horses. Some also try to protect wild animals from predation by other humans, or from the destruction of their habitat. Some campaign for the preservation of ecosystems. Some fight for the health of our planet, threatened by our own greed and consumption.

There is a limit to how kind anyone can be, and a limit to how many forms of suffering he or she can address. But all of us can be kinder than we are.

It is a basic spiritual practice to be kind within the context of one's own family and social groups, going beyond the minimum standards set by custom and morality. It is a more spiritual practice to be kind to people and other living beings beyond our own familiar circle.

We would not be here without the care of our parents and carers, or without those who cared for previous generations, and for prehuman ancestors through millions of generations. Our ancestors pass on to us the gift of life; and we pass it on to those who come after us. We are in a chain of giving. When we realise that our own life and everything is a gift, we are more inclined to give thanks, and to give. Kindness connects, and is itself a fruit of connection. It is part of the flow on which we all depend.

Spiritual practices may make us feel good, even blissful, deeply connected, and filled with love. But this connection is not an end in itself. It is about our interconnectedness with the unity underlying all being. Spiritual practices build up this sense of connection. It is easier to be kind when we feel ourselves linked to the ultimate source of joy, bliss and love. The practice of kindness comes more easily if it flows from a sense of fundamental connectedness and ultimate unity. Other spiritual practices, as discussed in this book, help to develop this sense of connectedness. The more connected we feel, the wider our recognition of kin, and the stronger our impulse to kindness. Spiritual practices can help us to be kinder.

Two practices of kindness

GIVING AWAY A PROPORTION OF YOUR INCOME OR WEALTH

All religious traditions encourage giving, and provide clear guidelines for doing so. In the Judeo-Christian tradition, giving away a tenth of one's income as a tithe was the usual standard.[60] The English word 'tithe' comes from the same root as 'tenth'. In Islam, there is also a kind of charitable giving called *zakat,* which is traditionally an annual gift of two and a half per cent of a person's surplus wealth.

Now we all pay government taxes, some of which fund traditional charitable causes like schools, hospitals, orphanages, and the care of widows, orphans, and old and sick people. Government taxes make contributions compulsory, and take away the voluntary or spiritual element. But at the same time, most tax systems also facilitate giving beyond the minimal demands of the law. Many countries have tax incentives for charitable giving.

If we want to make giving a spiritual practice, then it should be a regular practice, not only a caprice or whim. So it is helpful to think about a percentage of wealth or income or time as a target for giving. Then think about how this giving can be done most effectively and helpfully. And then give, and keep giving.

PRACTICAL KINDNESS

Monetary giving is the easiest way of being kind for those who are cash-rich and time-poor. Giving through personal services to others, and spending time with them, is much more immediate, and more directly loving. And it is good to give both money and direct help. There are many opportunities for helping others in our everyday lives, and also many charitable organisations through which we can volunteer to help other people, other species and the ecosystems on which all our lives depend.

Being kind literally means recognising our relationship or kin to other people, or other species, and ultimately to everything. All spiritual practices enhance our sense of connection with realities greater than ourselves. The stronger the connection we feel, the kinder we can be.

8

Why Do Spiritual Practices Work?

There are many kinds of spiritual practice. In the preceding chapters of this book, I discuss seven; in my previous book, *Science and Spiritual Practices*, I discuss another seven. There are still more, as I mention below. All these practices have measurable effects. In various ways they affect our physiology, breathing, heart rates, autonomic nervous systems, hormone levels, brain activities, mental abilities, feelings, emotions, visual imagery, sense of beauty, feelings of wellbeing, happiness, and compassion.

But how can such very different activities – like meditation and sports, fasting and chanting, taking psychedelics and participating in rituals – all have spiritual as well as physiological effects? Why do they work?

In our contemporary secular context, many people take up spiritual practices for their health benefits, or to become happier and more successful. But traditionally, in their original religious contexts, the beneficial effects of such practices were not the primary reason for doing them, but by-products of an underlying desire to come into conscious relationship with more-than-human forms of consciousness.

Through spiritual practices, many people feel a connection to a greater consciousness, or presence, or being. They often experience this connection as blissful or joyful. Even one brief experience of a state of blissful connection can be enough to change the course of someone's life. And such life-changing experiences can come spontaneously without any spiritual practices at all, as in near-death experiences and spontaneous mystical experiences.[1]

But do these experiences of connection really relate to forms of consciousness 'out there'? Do spirits, gods, and goddesses actually exist? Is there a conscious ground of being that underlies the universe, or an ultimate state of bliss that Buddhists call nirvana?

Or are all these experiences inside our bodies, and especially inside our brains? Is the experience of connecting with a consciousness greater than our own an illusion? Is it no more than an altered state of our own mind that is generated by unusual patterns of neuronal activity? As I discussed in the introduction to this book, most atheists and materialists answer 'yes' to these questions.

The usual materialist worldview is that we live in an unconscious universe. The only forms of consciousness are those that have emerged in complex brains, and above all in human brains. Other animals may be conscious to lesser degrees, dogs probably more than frogs; lower animals, like worms, have less mental activity, if they have any at all. Perhaps on other planets, in other solar systems, there are biological beings with brains analogous to ours – in other words, aliens or extraterrestrials. But there are no immaterial gods or spirits 'out there'. Spiritual practices work through their physiological, chemical and physical effects on bodies and brains, not by contacting mysterious spiritual beings outside the physical world.

Our brains are often assumed to be like computers. Some people speculate that computers themselves will eventually become conscious and even threaten our own existence by surpassing our intelligence. Some foresee a technological singularity in the more or less distant future, when computers with upgradable artificial intelligence will undergo a series of self-improvement cycles at an accelerating rate, causing an explosion of intelligence and giving rise to a powerful super-intelligence that far surpasses our own.

But this super-intelligence would still be inside machines. Could intelligent machines acquire consciousness, and even acquire the desire and ability to take part in spiritual practices? And could these practices connect them to spiritual realms? From a materialist point of view, this is a meaningless question, because there are no spiritual realms 'out there'.

Materialist spirituality

As we have seen, despite their belief that consciousness is confined to brains, many materialists take part in spiritual practices. How and why do they think these practices work?

They rightly point out that spiritual experiences are associated with changes in the body and brain. Taking part in sports depends on psychological and neuromuscular abilities – an ability to concentrate and coordinate movements skilfully. Psychedelic experiences depend on brain chemistry through modulating neurotransmitter systems. Meditation affects the activity of different regions of the brain as revealed in fMRI scans, and decreases activity in the default mode network associated with ruminations and worries (*SSP*, chapter 1).

Singing or chanting in groups can bring people into a literal resonance through their sounds, and also through synchronised breathing (*SSP*, chapter 5). Fasting activates the use of fat reserves, leading to increased levels of ketones in the blood, affecting the activity of the brain. In general, spiritual practices have physiological effects. They influence moods, emotions, brain activity and wellbeing.

Mystical experiences can occur spontaneously in connection with temporal lobe epilepsy. The Russian writer Fyodor Dostoyevsky had such experiences himself and described them as a 'happiness unthinkable in the normal state and unimaginable for anyone who hasn't experienced it . . . I am in perfect harmony with myself and the entire universe.'[2]

Religious people generally have no problem in accepting that spiritual experiences are associated with changes within bodies, but they do not see these physical changes as the whole story.

Whereas materialists see the bodies and brains as the producers of consciousness, religious people generally see them as more like filters or windows. Most religions explicitly recognise forms of consciousness beyond the human level: gods, goddesses, spirits, angels and God. They take it for granted that these spirits can communicate with humans through a range of conscious experiences, as well as through bodily effects, such as healing, and also by affecting the course of events.

The borderline case is secular Buddhism, as distinct from Buddhism as it is actually practised in the traditional Buddhist countries of Asia. Secular Buddhism is Buddhism reinterpreted in the light of materialist science, with its mind-in-the-brain assumption. Some sophisticated contemporary atheists are secular Buddhists, like Sam Harris[3] and Susan Blackmore.[4] They meditate. They follow essential Buddhist principles. But they see Buddhism as a kind of psychology that illuminates the nature of the mind, rather than as a religion.

They differ from traditional Buddhists, like Theravada Buddhists in Sri Lanka, Zen Buddhists in Japan, and Vajrayana Buddhists in Tibet, who have prayers, festivals, temples, and pilgrimages, and also a belief in rebirth. The current Dalai Lama, for example, is believed to be the fourteenth Dalai Lama in a succession of incarnations, with gaps of only a few years in between the death of one and the birth of the next.

In the *Dzogchen* tradition of Tibetan Buddhism, the ultimate goal is the achievement of the rainbow body. When a highly accomplished practitioner dies, his or her body disappears into a rainbow of light, leaving only the hair and the fingernails behind.[5] And many Buddhists believe in a series of Buddhas, each incarnating a form of Buddha consciousness appropriate for his time and age. They believe that there were already twenty-seven Buddhas before Gautama Buddha – the historical Buddha – and another Buddha, Maitreya, is yet to come.

Secular Buddhists reject all these aspects of Buddhism.[6] They reinterpret Buddhism as a form of atheism or secular humanism.[7] And indeed Buddhism does not have a creator god. Buddhism has very little cosmology at all, preferring to focus on personal transformation rather than discussing the essential qualities of nature, or the existence of God. This was brought home to me very clearly when I went to visit a revered Tibetan lama, Lama Wangdor, in his cave in the Himalayas, above the Lotus Lake of Tso Pema, in Himachal Pradesh, India.

I was intrigued by the idea that the sun is conscious, and that all other stars might be conscious beings. I wanted to know what he thought about this. I climbed a rocky path to his cave. He received me courteously, listened patiently to my questions, was open to the idea of stellar consciousness as a possibility, but was not very interested. He said, 'What is important is that you practise yourself.' His message was: meditate, rather than waste your time speculating about nature.

The Buddha Gautama himself had a similar attitude. He avoided speculation about the natural world as a distraction from seeking liberation through the noble eightfold path, and above all through the practice of meditation. Nevertheless, traditional forms of Buddhism are very different from modern secular Buddhism. Unlike the materialist philosophy, Buddhism does not assert the primary reality of non-conscious matter, but an ultimate ineffable state of consciousness, nirvana. In so far as nirvana is the greatest bliss or happiness,[8] it is conscious, not non-conscious.

Secular Buddhists argue that religious forms of Buddhism require an act of faith, and that this is unnecessary. But to the extent that secular Buddhists put their trust in materialism, this is itself an act of faith. Materialism is not the Truth; it is a belief system. From the materialist point of view, minds are what brains do. Hence, for materialists, when people have altered states of consciousness, psychedelic hallucinations, epilepsy-induced states of bliss, or spontaneous mystical experiences, these are all a result of changes in the brain. Materialists assume that spiritual practices cannot connect us to more-than-human forms of consciousness. They believe that more-than-human forms of consciousness do not exist.

Connecting with more-than-human consciousness

Most people focus their attention on the differences between religions. Historically these differences have led to many conflicts and disputes. But instead, if we look at the similarities, they are very striking. Here I attempt to summarise these fundamental doctrines because they can help us think about spiritual practices and why they work.

All religions presuppose realms of consciousness beyond the human level.

If we live in a universe that is fundamentally conscious rather than unconscious, then what is the nature of this fundamental consciousness?

By definition, an all-embracing consciousness is inconceivably greater than our own. In mystical experiences, we may connect directly with this greater consciousness. But when it comes to mental models or theories framed in human conceptions and languages, our imaginations are almost certainly too limited to embrace the very nature of ultimate reality.

Religions based on deep mystical experiences see the ultimate reality as being entirely beyond any conception we can form, and definable only in negative terms. We can only say what this ultimate reality is not, rather than what it is. Within Christianity, this approach is called apophatic or negative theology, and is particularly widespread in the Eastern Orthodox tradition. The German mystic Meister Eckhart (c. 1260 – c. 1328) called this ultimate reality the Godhead, beyond all mental forms, qualities or descriptions.

In India, the ultimate Absolute Reality, Brahman, is described negatively, or apophatically, as Brahman without attributes or qualities, called Nirguna Brahman.

In Buddhism, this ultimate state is defined negatively as *nirvana*, sometimes translated as 'the void' or 'emptiness', and more literally as 'extinction' or 'disappearance'. The original meaning of the word is 'blown out'. It refers to the liberation of humans from the process of rebirth. It is associated with a profound peace that goes beyond all forms of suffering.

Buddhism grew out of a world-renouncing ascetic movement in Northern India around 500 BC. Its goal was to put to an end the cycles of rebirth, eradicating all karmic traces, which were seen as the source of suffering. The ultimate aim was 'the cessation of personal identity, in a kind of permanent ontological suicide (i.e. the irreversible destruction of one's very being).'[9]

Some forms of spiritual practice, particularly meditation, can lead to the experience of connection with this ultimate conscious reality, surpassing all normal human limits and mental constructs, and entering a state of bliss inherent in the very nature of the ultimate. But most spiritual practices do not reach this ultimate level. They are more concerned with our activities in the world. For example, petitionary prayer is not about ultimate realities, but about worldly needs and goals.

All religions agree that the ultimate reality is conscious, with a far greater consciousness than ours, unimaginably beyond our limited human conceptions. And yet this ultimate reality is related to the world in which we live, and to our own minds and societies. What is the connection between ultimate consciousness and nature?

The threefold nature of God

The ultimate reality is indescribable, without qualities – nirvana, the Godhead, Nirguna Brahman. But the interface between this ultimate reality and the natural world is widely thought of as threefold or trinitarian.

The threefold nature of God is reflected in nature itself, and in human life and minds. Our minds reflect the divine mind. To use a modern mathematical metaphor, they are fractals of the ultimate mind.

According to the *Kena Upanishad*, one of the foundational holy books in the Hindu tradition, Brahman, God, is not an object. God is not something that the eyes can see, or the ears can hear, a thing among other things. Instead, Brahman is that by which the mind comprehends, by which the eye sees, by which the ear hears. Atman, which is at the centre of our own conscious being, is the eye of the eye, the ear of the ear, and the ground of all knowing. Our minds participate in the mind or knowing of Brahman through knowing themselves: 'What cannot be seen with the eye, but that whereby the eye can see: know that alone to be Brahman.'[10]

In the Hindu conception, Saguna Brahman, Brahman with qualities, God as manifested in the world, has three fundamental aspects *sat*, *chit* and *ananda*, being, consciousness and bliss, *sat-chit-ananda*. As Bede Griffiths put it, 'God, or Ultimate Reality, is experienced as absolute being (*sat*), known in pure consciousness (*cit*), communicating absolute bliss (*ananda*). This was the experience of the seers of the Upanishads as it has been that of innumerable holy men in India ever since. It is an experience of self-transcendence, which gives an intuitive insight into Reality.'[11]

Different schools of Indian thought have their own versions of this trinity. In Kashmiri Shaivism, the ground of all being is called *Parashiva*; the source of form and order is *Shiva* and the primordial energy of the cosmos is *Shakti*. Shakti is feminine.

Within the Christian tradition, God is the Holy Trinity: Father, Son and Holy Spirit. The ground of all being is the Father. The Son is the Word or Logos, the source of form and order; the Spirit is the breath, wind, or energy, like Shakti. In Indian languages, the words for spirit – *ruach* in the Hebrew of the Old Testament and *pneuma* in the Greek of the New Testament – become Shakti. Like Shakti, ruach is feminine. All three aspects of the Holy Trinity constitute the unity of God acting in and through the universe, creating and sustaining it.

In the Judeo-Christian tradition, the Word of God is primarily the spoken word, not the written word. Speaking is the fundamental metaphor for the Christian conception of God as Holy Trinity. Spoken words involve on the one hand structures, forms, patterns and meanings, and on the other hand a flow of breath or spirit. Think of your own speech. You are the being on which both your words and your breath depend and from which they come forth.

In this metaphor, you are the ground of Being. If there are only words in your mind, unspoken, they remain latent. They are unmanifested Logos. If there is an outflow of breath and no words, there is merely an energetic flow, with no form or meaning, Spirit without form. But when the form and order of your words are carried on your outbreath, your words can communicate and connect. The Logos and Spirit work together.

Although we can distinguish between the speaker (the ground of Being), the words that are spoken (the Logos) and the breath on which they are carried (the Spirit), in speech all three aspects have an underlying unity. They are three in one and one in three.

In some Christian interpretations of the Holy Trinity, like St Augustine's psychological model, God the Father is the knower, God the Son or Logos the known, and the Holy Spirit is the joyful love between them. This conception is very similar to *sat-chit-ananda*.

In both the Hindu and Christian traditions there is an ambiguity about the dynamical principle, namely Shakti or Spirit. On the one hand she is bliss, joy and love, and on the other hand she is the principle of movement or change – creative and destructive power, cosmic energy, breath, wind and life-spirit. What is the relationship between bliss and energy, which are both aspects of Shakti or Spirit?

Divine consciousness is essentially blissful, and this blissful aspect, *ananda*, is distinguishable from the ground of being, *sat*, and knowledge or consciousness, *chit*. The ultimate conscious being, Brahman or God, is full, not in a state of need or lack or desire. And yet this is not a static bliss or joy beyond all movement and change, but the basis of all movement and change. As Griffiths put it, the *ananda*, the bliss or joy of the Godhead, is 'the outpouring of the superabundant being and consciousness of the eternal, the Love which unites Father and Son in the non-dual Being of the Spirit.'[12]

In the Muslim tradition, God's oneness likewise includes being, consciousness, and bliss, which are called *wujud*, *wijdan* and *wajd*.

All these traditions share much common ground. As the theologian David Bentley Hart summarises it:

In God, the fullness of being is also a perfect act of infinite consciousness that, wholly possessing the truth of being in itself, forever finds its consummation in boundless delight. The Father knows his own essence perfectly in the mirror of the Logos and rejoices in the Spirit who is the 'bond of love' or 'bond of glory' in which divine being and divine consciousness are perfectly joined. God's *wujud* is also his *wijdan* – his infinite being is infinite consciousness – in the unity of his *wajd*, the bliss of perfect enjoyment. The divine *sat* is always also the divine *chit*, and the perfect coincidence is the divine *ananda* . . . God is the one act of being, consciousness, and bliss in whom everything lives and moves and has its being; and so the only way to know the truth of things is, necessarily, the way of bliss.[13]

Chinese philosophy starts from very different principles, but the Taoist conception of the polarity of yin and yang interacting in all nature is also trinitarian. In the familiar symbol of their relationship, yin, the dark swirl, contains within it a seed of yang, the light swirl, and vice versa. Their polarity is not a dualistic opposition, but rather an interdependence or complementarity. Both yin and yang are part of an ultimate unity, the Tao, which includes them both, symbolised by the circle that contains the interlocking swirls.

Scientific trinities

At first sight, these theological discussions seem very remote from the secular world of modern science. But a trinitarian model underlies much scientific thinking, too. The laws of nature play the role of the Logos, the principles of form and order, and energy is the Spirit principle, the basis of all movement, change and activity.

When the mechanistic theory of the universe first came into being in the seventeenth century, it was grounded in a traditional theology in which God was both the source of order in the natural world, and the source of human reason, through which this order could be understood. God's Logos included the laws of nature, which were principally mathematical, and this divine Logos was also the source of humanity's God-given reason.

This view was originally part of a much richer picture. The theory of the primacy of 'the laws of nature' emphasised only one aspect of the Logos, and froze it into static laws, and left out its traditional creative and imaginative aspects. Nevertheless, it still connected human reason with the source of nature itself.

The theology of many Enlightenment philosophers, including Voltaire, was deism, a belief in a God of Reason. The triumphs of scientific understanding – like Newton's law of gravitation –

seemed to show that human reason could participate in the Reason that underlay the natural world. Humans were becoming godlike through their reasoning, through mechanistic science, through technology and through economic progress.

The Romantics reacted against this one-sided intellectual understanding by emphasising change, flow, creativity, emotion – in other words, the dimension of the Spirit. And around the same time, in the early nineteenth century, the concept of energy became a grand unifying principle for physics. Progress needed energy, both literally and metaphorically. Literally it came from flowing wind and water in windmills and watermills; later through burning coal; then through oil and natural gas; then nuclear power; and now once again through renewable sources, including water, wind and solar power.

But energy cannot by itself give form or organisation. This was the role of fields, introduced into science by Michael Faraday in the 1850s in the concept of electrical and magnetic fields with their geometrical lines of force. Fields are spatial regions of influence.

In the twentieth century, the field concept was extended to the entire universe, through Einstein's model of the universal gravitational field, which is not *in* space and time, but *is* space-time, the context in which everything happens. Current models of cosmology include formative principles – fields – and dynamical principles – energy – and they also implicitly presuppose a ground of being from which these come and which sustains them. Fields were also extended into the very heart of matter through quantum field theory. An electron is a quantum of energetic vibration in an electron field. A proton is a quantum of energetic vibration in a proton field.

In biology, the concept of form shaping, or *morphogenetic*, fields was introduced in the 1920s.[14] These hypothetical fields shape developing organisms. For example, a growing oak seedling is shaped by an oak morphogenetic field, which guides its development towards the mature form of an oak. Likewise, peacock morphogenetic fields shape developing peacock embryos in peacock eggs.

These fields are invisible patterns of influence that mould the processes of development. They do so through a hierarchy of sub-fields. The overall oak field underlies the polarity of root and shoot, and includes all parts of the plant. It coordinates the fields of different organs, like roots, leaves and flowers, and these fields in turn coordinate the fields of tissues, which coordinate those of cells.

Among biologists, there is no general agreement on the nature of these fields. Some, including me, think that these are new kinds of fields, not yet recognised by physics. Others think that morphogenetic fields are simply words that are placeholders for detailed explanations that scientists will eventually be able to formulate in terms of regular physical fields, like quantum matter fields and electromagnetic fields. But even if morphogenetic fields could be reduced to the known fields of physics, the development of living organisms would still be organised by fields.

Fields are the formative principles within nature, and energy the principle of movement and activity. Fields and energy can be seen as physical analogues of Logos and Spirit, and like Logos and Spirit, they are intertwined.

Old-school materialism regarded matter as the ultimate, primal reality underlying all things. But that view is out of date. As the philosopher of science, Karl Popper, put it, 'Through modern physics, materialism has transcended itself.'[15] Matter is no longer fundamental. Fields and energy are more fundamental. Particles such as electrons and protons are not hard enduring stuff, like little billiard balls; they are vibratory structures of activity. They are made up of energy bound within fields. Energy gives things their actuality, their activity and their ability to interact, and fields give them their shape, form and organisation.

The sun, for example, is an energetic body. It emits huge amounts of energy. At the same time, its bodily shape – roughly spherical – is a result of its gravitational field. The patterns of energy within its body, and on its surface in sunspots and magnetic field lines, are manifestations of its complex, pulsating electromagnetic fields.

In roughly eleven-year cycles, its overall magnetic field reverses, and within this overall cycle it has faster and faster rhythms within rhythms. The solar wind emitted by the sun, in which the earth is bathed, has a fractal structure, with rhythmic patterns within patterns within patterns . . . [16] But these patterns are unpredictable from first principles, like the weather in earth, which is why there are regular solar weather forecasts, available online.[17]

All self-organising systems share in this interplay of energy and fields, including molecules, crystals, microbes, plants, animals, societies, ecosystems, planets, solar systems and galaxies. All of them are made up of flows of energy patterned by fields.

So are our own spoken words, the source of the metaphor of Logos and Spirit. The scientific account of speech adds detail to this traditional metaphor. We exist – we are beings – and we participate in the ground of Being by our very existence. When we speak, the outflow of our breath provides the energy for our speech, the Spirit, setting up acoustic vibrations in the air around us. Complex electromagnetic and electrochemical patterns of activity within our muscle cells, nervous systems, sensory organs and brains shape the movements of our tongues, lips and vocal cavities.

These are aspects of the Logos, and the shaping of the flow of breath produces the sequences of words. The sequences of words are themselves organised into meaningful patterns that connect us with each other. Words are a manifestation of the formative (Logos) principle inherent in minds, mediated through electromagnetic fields, and at the same time they depend on the flow of energy in nerve impulses and the flow of breath on which the spoken words are carried (the Spirit).

The same principles apply to modern cosmology. Form and order (Logos) are not sufficient. There has to be energy, too (Spirit), and a ground of Being that underlies them both. At the end of his best-known book, *A Brief History of Time*, Stephen Hawking discussed a possible mathematical theory of everything that would provide a scientific description of the universe, but admitted that such a theory would not be enough:

> Even if there is only one possible unified theory, it is just a set of rules and equations. What is it that breathes fire into the equations and makes a universe for them to describe? The usual approach of science of constructing a mathematical model cannot answer the questions of why there should be a universe for the model to describe. Why does the universe go to all the bother of existing?[18]

In theological terms, both the Logos and the Spirit depend on a ground of Being that sustains them both. The Logos alone could not create a universe and sustain its existence. Or, in scientific language, the laws of nature without energy could not make anything exist. Both the laws and the energy need to *exist*, to have being. And energy is needed to 'breathe fire into the equations'.

In his book, *A Universe from Nothing: Why There is Something Rather Than Nothing* (2012), the physicist Lawrence Krauss attempts to provide an atheist cosmology. He wants to get rid of God. To do so, he assumes that laws of nature exist and govern the universe from the beginning; he presupposes laws that transcend the universe, and even a multiplicity of laws that might govern other universes as well. To breathe fire into the equations, he postulates the existence of a primordial quantum vacuum that contains a vast 'energy of empty space', in which fluctuations can give rise to universes. But this empty space is not nothing; it exists as a 'boiling brew of virtual particles that pop in and out of existence in a time so short we cannot see them directly.'[19]

Thus, in order to conjure his universe from nothing, Krauss presupposes not nothing, but the existence of laws and energy before the universe came into being. His model is very similar to traditional trinitarian theology, except that he assumes that the ultimate source of everything is non-conscious and joyless, rather than conscious and joyful.

This is where spiritual experiences make a difference. Both materialist philosophy and theological doctrines are abstract and intellectual. They can be learned from books. But our under-standing of consciousness is much deeper when it is based on direct experience. Consciousness *is*, after all, direct experience.

We come back to the question of the nature of consciousness, and the question of consciousness beyond the human level.

Materialists are almost inevitably atheists, because the materialist philosophy sees the basis of all reality as non-conscious matter, in a way that leaves no scope for divine being or consciousness. They exclude consciousness from ultimate reality by definition, following Descartes, who defined matter as non-conscious in the seventeenth century. Descartes' dualistic worldview also included a realm of non-material consciousness that included God, angels and human minds.

Materialism is based on a rejection of this duality, abolishing the realm of immaterial consciousness and leaving only unconscious matter. This philosophy is literally godless. As we have seen, it also creates the 'hard problem' of trying to explain why humans are conscious in terms of non-conscious brain processes.

But not all atheists are materialists. Some see nature as alive, and pervaded by mind – but it is the mind of nature, rather than of God. For example, the poet Percy Shelley (1792–1822), a pioneering romantic atheist, was convinced of the reality of a living power in nature, which he called the 'Soul of the universe' or the 'all-sufficing Power' or the 'Spirit of Nature'.[20] The children's writer Philip Pullman had a mystical experience as a young man, which convinced him that the universe is 'alive, conscious and full of purpose.'[21] He says of himself, 'I'm religious, but I'm an atheist.'[22]

Some people who are committed to the materialist worldview feel a need to enlarge or soften it, because they have experienced a deep conscious connection with the more-than-human world (*SSP*, chapter 3). Sometimes these transformative experiences come about through psychedelics. For instance, the science writer Michael Pollan self-identifies as a materialist, but his psychedelic explorations have taken him far beyond the standard materialist theory:

> One of the gifts of psychedelics is the way they reanimate the world, as if they were distributing the blessings of consciousness more widely and evenly over the landscape, in the process breaking the human monopoly on subjectivity that we moderns take as a given. To us, we are the world's only conscious subjects, with the rest of creation made up of objects. Psychedelic consciousness overturns that view, by granting us a wider, more generous lens through which we can glimpse the subject-hood – the spirit! – of everything, animal, vegetable, even mineral, all of it now somehow returning our gaze. Spirits, it seems, are everywhere.[23]

The philosophy of materialism has a continual tendency to open up to animism, or panpsychism, through recognising life and mind in the natural world.

Animism or panpsychism

Animism is both an archaic view of nature, and a very modern one. It preceded the machine theory of nature, which has dominated the sciences for four hundred years. In response to the 'hard problem' of explaining why humans are conscious, increasing numbers of philosophers and neuroscientists are adopting the philosophy of panpsychism – the view that there is some kind of mind or consciousness in all self-organising systems, including electrons and atoms.[24] Thus the emergence of human minds within human brains and bodies involves a difference of degree, but not of kind, from other forms of matter.

Panpsychism is, in effect, a modern form of animism. The mechanistic paradigm is also breaking down with the realisation that nature is made up of organisms rather than inanimate mechanisms. Indeed, the entire universe now looks like an organism, because of its evolutionary, self-organising nature. At the same time, many people's personal relationships to nature are being transformed through neo-shamanism and psychedelic experiences.

As recently as 1966, when the Big Bang theory became predominant, the orthodoxy of physics still saw the universe as an eternal machine. But since then, cosmologists have come to see the entire universe as expanding and evolving from its very small beginnings. This new evolutionary cosmology recalls the ancient myth that the universe began with the hatching of a cosmic egg.[25] The universe is more like a developing organism than a machine.

Traditional animism is, by definition, a belief that many aspects of nature are animated by souls – in Latin, the word for soul is *anima*. In the animism of ancient Greece, as codified by the philosopher Aristotle, souls were the animating principles of all living beings, and their primary role was formative: they attracted developing organisms towards their characteristic form. The soul of a cedar tree gave the cedar tree its form. The soul of a gazelle gave the gazelle its form, and its instincts.

In this view, the natural world depends on both formative and energetic principles, which in later theology were seen as derived from Logos and Spirit.

The formative principles were souls, and what we now call the energetic principle was *prima materia*, or prime matter, in itself formless, and pure potentiality. If there were only two principles, it would by definition be a dualism. But at least implicitly, soul-based animism is trinitarian. The formative principles of all living beings and the energy that makes them alive have a common source. These principles exist in such a way that they work together.

In the Middle Ages, the philosophy of nature, taught in the universities and monasteries of Europe, accepted that nature was permeated with life. It was not mechanical and inanimate. God was the God of a living world, and human minds participated in the divine mind. One of the principal architects of this intellectual synthesis was St Thomas Aquinas (1225–74). For Aquinas, following Aristotle, all living beings were animate or ensouled, including all plants and all animals: the soul was 'the first principle of life'.[26]

St Francis of Assisi (c. 1182–1226) wrote of 'our Sister, Mother Earth, who sustains and governs us'. His disciple St Bonaventure tells us that 'he would call creatures, no matter how small, by the name of "brother" or "sister".'[27] Humans differed from other animals not because they had souls, but because they had *rational* souls, and hence minds that reflected the mind of God.

As the philosopher, Joseph Milne, summarised it, 'For Aquinas, the act of creative knowing in the mind of God that brings all things into being is in a sense mirrored in the human mind in which all things are potentially received . . . Through the desire to know created things, the mind of man is drawn to know them as they are known in the mind of God, and to know the knower of all things. This beatific vision is the true mystical end of human knowing. In this act of mystical knowing, the creation comes into a mystical knowledge of itself.'[28]

With the mechanistic revolution in science in the seventeenth century, nature became inanimate machinery, and God became like an immaterial engineer that created the machinery in the first place and then left it to run automatically. God was no longer within nature, but outside nature. This machine-making God soon became an optional extra for mechanistic science, and then with the rise of the materialist philosophy became redundant.[29]

Materialism went hand in hand with atheism. Now neither nature nor human minds participated in divine being. Nature was unconscious and purposeless, and human minds nothing but physical activities inside brains that had evolved through mindless processes of natural selection.

However, in the nineteenth century, as discussed above, starting with the work of Michael Faraday, formative principles were reinstated within the natural world. They are no longer called souls but fields. And fields provide the best starting point for a new understanding of mind in nature, or psyche in nature, or panpsychism. They have a formative influence on energy, organising it and patterning its otherwise indeterminate activity. They are inherently holistic.

According to the hypothesis of formative causation, fields are shaped by an inherent memory given by the process of morphic resonance, and they underlie the formation of molecules, crystals, living organisms, solar systems and galaxies; they organise the activity of the nervous system, animal behaviour, instincts and learning, and the activity of human and animal minds.[30]

Panpsychism or animism sees nature as alive, and mind inherent in all kinds of organisms at all levels of complexity, from subatomic particles to the entire universe. Taken to its logical conclusion, it implies that there is a kind of cosmic or universal mind immanent in the universe. In this sense it resembles the seventeenth-century philosophy of Baruch Spinoza (1632–77), who saw nature as the body of God and God as the mind of nature. This view is usually called pantheism. God is nature and nature is God. God is synonymous with nature; there is nothing beyond nature.

Pantheism differs from supernatural theism, which supposes that God is outside nature, transcendent but not immanent. For the same reason, pantheism differs from deism, the doctrine of a transcendent creator God who set up the rational laws of nature, and started off the universe in the first place, but then left it to run automatically without further intervention.

Panentheism

Christian theology before the seventeenth century was much more inclusive. God was both immanent in nature, within nature, and transcendent, beyond nature. A new version of this traditional view is called *panentheism*. Panentheism takes an animistic or panpsychist view of nature, but it goes further. God is not only in nature, as in pantheism, or only transcendent, as in supernatural theism, but both.

Panentheism enables panpsychism to be related to theologies in which the natural world is a reflection of Being, Logos and Spirit, or *sat-chit-ananda*. God is in nature and nature is in God. God is both immanent and transcendent. This word resembles pantheism, which sees God everywhere, includes the Greek word *en*, 'in', and means 'in' in two senses: God in nature and nature in God. As Matthew Fox puts it:

Healthy mysticism is panentheistic. This means that it is not theistic, which envisages divinity 'out there' or even 'in here' in a dualistic manner that separates creation from divinity. Panentheism means 'all things in God and God in all things'. This is the way mystics envisage the relationship of world, self, and God. Mechthild of Magdeburg [c. 1210–c. 1285], for example, says, 'The day of my spiritual awakening was the day I saw and knew I saw all things in God and God in all things.' . . . Panentheism melts the dualism of inside and outside – like fish in water and the water in the fish, creation is in God and God is in creation.[31]

Fox says he has met many serious spiritual seekers who describe themselves as atheists. 'Yet I have come to realise that most atheism is a rejection of theism and most atheists are persons who have never had panentheism or mysticism named for them, in a culture where theistic relations to divinity are celebrated at the expense of mystical or panentheistic ones. I do believe that if the only option I was given by which to envisage creation's relationship to divinity was theism, then I would be an atheist, too.'[32]

I agree with Fox. This corresponds with my own experience. I actually was an atheist when I thought that we lived in a mechanical, non-conscious natural world and that the only religious option was a transcendent God completely outside the natural world, and outside normal human life. My views changed first through recognising the life of nature, and then through an appreciation of God's immanence within nature – within the forms of all living beings, and within the wind, breath and sunlight – and within myself.

Trinities and spiritual practices

Trinitarian models, explicit in Hinduism and in Christianity, and implicit in the mystical theology of Islam, provide a way of interpreting why very different spiritual practices provide spiritual connections to ultimate reality and are at the same time joyful. Without such models, it would be difficult to see how practices as different as meditation, chanting and sports could all have a spiritual dimension.

Meditation, which typically involves minimal physical activity, seems to be a way of connecting with the ground of being, *sat*, or the Father. The deepest forms of meditation can go further, beyond all differentiation within the divine, to the Godhead, or Nirguna Brahman or nirvana. This ultimate reality is blissful by its very nature. Participating in it gives the deepest possible joy.

Sports, by contrast, are not about stillness, but about movements directed towards a goal. The spiritual experiences that come through sports link more to the principle of flow, to Shakti, or spirit. They connect with the bliss of this flow. Likewise, singing, chanting, dancing and music connect with the flow of the spirit and the joy that accompanies it. Watching the graceful movements of animals also shows us a combination of form and energy coming from a common source.

By contrast again, the contemplation of the beauty of flowers (*SSP*, chapter 4) and other experiences of visual beauty have less to do with flow and more to do with form, or idea, or *chit*, or Logos.

Fasting is not in itself a spiritual experience, but is a practice that interrupts the normal habits of appetite and bodily desire. It creates a space in which spiritual realities can be more present. The decision to fast is taken with the intention of going beyond regular desires and habits. Fasting creates a mental and physical context in which other spiritual practices, like prayer and meditation, can be more effective.

Holy days and festivals create spaces in which regular spiritual practices like prayer, chanting, singing and rituals can be the principal focus of activity, as opposed to work. These celebrations bind communities together, relating them to the cycles of the more-than-human world, and to the ultimate source of nature and humanity.

Prayer provides a way of explicitly linking our own minds, needs, fears and intentions to the greater minds of the spiritual beings to whom we pray. Petitionary prayer links us to the flow of events, and offers us the possibility of being co-creators of what happens, rather than passive recipients.

Some psychedelic experiences, especially with substances such as DMT and 5-methoxy DMT, take the experiencer to what seems like the ground of being itself, or *sat*. But most psychedelic experiences create an intense immersion in the realm of imagination with its ever-changing forms and meanings.

As Terence McKenna put it, our minds have an affinity for minds and for the order that minds produce: 'All pattern seems to quickly lose its charm unless it's pattern that has been put through the sieve of the mind . . . We look for an aesthetic order, and when we find that, then we have this reciprocal sense of recognition and transcendence, and this is what the psychedelic experience provides in spades.'[33]

Psychedelic visions are combinations of form and energy in the world of the imagination, Logos and Spirit. And there may be many imaginations, not just our own individual imaginations. We may all participate in a collective human imagination, expressed through archetypal forms in our dreams, fantasies and visions, which the psychologist C.G. Jung called the collective unconscious, shaped by collective memories.[34]

Other species may also have imaginations that work through their dreams. We cannot ask animals what they dream about, but when they are sleeping, they show physiological changes, like rapid eye movements, very similar to dreaming humans. According to the *Oxford Companion to Animal Behaviour*, 'On the basis of the evidence, many scientists are willing to agree that many animals experience dreams that are akin to those of human beings.'[35]

For example, dogs dream, and probably dream about things that dogs can do, and perhaps about things they cannot do, like fly. Perhaps the dream worlds of different species sometimes overlap and influence each other. Maybe the entire planet, Gaia, has a dreamlike imagination; maybe the sun has a solar imagination; maybe the galaxy has a galactic imagination; maybe the entire cosmos has a cosmic imagination. And all these imaginations may be derived from and within the divine imagination, which contains all possible forms, ideas, words, meaning, experiences, and scenarios.

In dreams and in psychedelic experiences, we may not be confined to the imaginal realm of human minds, but contact the imaginal realms of other species, of the earth and the heavenly bodies, and ultimately of the divine mind in its joy.

In so far as all spiritual practices can lead us towards a greater sense of connection with the whole, or the All, or the love of God, then they expand our awareness of our kinship with other people, with other animals, with plants, with the earth, and with all nature. They motivate us to behave more kindly, and to live and work for the greater good.

Evolutionary spirituality

In the realm of spirituality, some of the greatest breakthroughs occurred many centuries ago, including discoveries about the nature of consciousness made by shamanic explorers in hunter-gatherer societies, the Indian rishis or seers, the Buddha, the Hebrew prophets, the visionary philosophy of Plato, the experiences of Jesus, and the insights of medieval Christian and Sufi mystics.

By contrast, the findings of the natural sciences are much more recent, and scarcely a month passes without the announcement of some new scientific or medical discovery. This contrast can easily lead to the superficial conclusion that the sciences are all about discovery, progress and looking forwards, whereas religious and spiritual traditions are fixed, and are about looking backwards.

It is true that all religions honour their traditions and their ancient roots. But scientists also look back to the founding fathers and heroes of the scientific quest, like Copernicus, Galileo, Kepler, Descartes, Newton, Faraday, Darwin and Einstein. And just as the sciences are in a process of continual development, so are religion and spirituality.

In fact, in the twenty-first century, religions and spiritual practices are evolving at an unprecedented rate. The social and economic contexts in which they are practised have changed beyond recognition through modern technologies, the internet, the global economic system, travel, migration, and modern education.

The meditational techniques practised by Buddhist monks for centuries have been propagated throughout the world by teachers of mindfulness meditation and through the diaspora of Tibetan lamas and monks. Likewise, the transmission of transcendental meditation and other meditation techniques from India has reached hundreds of millions of people globally.[36] Yoga is practised by many millions of people far from its ancestral home in India, and is rapidly evolving into a host of different methods. The traditions of oriental martial arts are now taught in many different countries and are undergoing their own evolution, including a kind of natural selection through mixed martial arts, in which different techniques are pitted against each other.

Shamanic practices, once particular to some tribal peoples, dismissed as primitive animists by most Europeans and North Americans in the nineteenth century, are now being taught in workshops throughout the Western world, and thousands of people go to Peru and other countries in South America for experiences of psychedelics in a shamanic context.

Meanwhile, the global spread of Christianity has led to its transformation in Africa, Asia and South America as it becomes assimilated to indigenous cultures. Pentecostalism, with its emphasis on gifts of the Holy Spirit, traces its roots to the early church, but emerged in a modern form in the early twentieth century in Los Angeles. It started as a radical evangelical movement that included the practices of speaking in tongues and spiritual healing, and has now been carried worldwide through Pentecostalist churches and through the Charismatic movement within established churches.

Today there are about 580 million Pentecostalist and Charismatic Christians, making up about a quarter of the total number of Christians in the world, many of whom are in previously non-Christian countries.[37]

In Brazil and the Southern United States, a combination of Christian teachings and indigenous psychedelic practices has led to the emergence of psychedelic churches, like the Native American Church, based on the visionary mescaline-containing cactus peyote, and the Brazilian ayahuasca-based churches of Santo Daime and União do Vegetal.

At the same time, in the secular world, there has been an extraordinary evolution of unitive 'in-the-flow' experiences through sports, including many new kinds of sport that take human experiences into previously uncharted territory.

Ecstatic dancing in raves, clubs and festivals is creating new kinds of collective experience, heightened and transformed through the use of substances like MDMA, or Ecstasy.

With the development of new psychoactive substances like LSD, many millions of people have explored realms of consciousness in contexts very different from the traditional uses of psychedelics.

Most of these evolutionary developments have taken place outside the framework of established religions, whose traditions did not prepare them for such rapid changes. They have also taken place beyond the purview of institutional science, which because of its secular ideology and anti-spiritual bias has largely ignored or dismissed them until recently. They have also occurred outside the boundaries of official educational systems, which confine themselves to academic activities that have little to do with direct experience.

But change is in the air. As I have tried to show in this book, as well as in *Science and Spiritual Practices*, we are entering a new phase of spiritual and scientific evolution. Scientific studies of spiritual practices are illuminating how they affect our bodies, brains, health, wellbeing and behaviour. These studies may in turn help to make these practices more effective, and enable them to be taught and transmitted better. And in addition to the seven spiritual practices I have discussed in this book, and the seven I discussed in *Science and Spiritual Practices*, there are many more, including yoga, chi gong, tai chi, caring for the dying, confession, devotional worship or *bhakti,* forgiveness, dream yoga, dream incubation and art.

To take just one of these examples, dream incubation is an ancient spiritual practice that could relatively easily be revived, studied scientifically, and reintegrated into evolving religious traditions. In several ancient cultures, people went to holy places where they slept with the intention of receiving healing or oracular dreams. In ancient Egypt, one of the most important places for oracular dreaming was in the temple of Seti at Abydos.[38]

As described in the Old Testament, King Solomon slept in a holy place after offering sacrifices to God, and God appeared to him in a dream: 'The king went to Gibeon to sacrifice there, for that was the principal high place; Solomon used to offer a thousand burnt offerings on that altar. At Gibeon the Lord appeared to Solomon in a dream by night; and God said, "Ask what I should give you."' (1 Kings 3: 5 –6.)[39] Solomon asked for, and was given, 'an understanding heart'.[40] The legendary wisdom of Solomon was rooted in a dream in a holy place.

In ancient Greece, pilgrims went to the temples of the healing god Asclepius at Epidaurus and at other centres. After undergoing ritual purifications and making offerings to the god, they slept within the consecrated space with the intention of experiencing healing or oracular dreams. Priests helped them to interpret their dreams and visions. Dream incubation was also practised in temples throughout the Roman Empire, including Britain,[41] and this practice continues within the Greek Orthodox Church, especially in churches dedicated to the twin healing Saints Cosmas and Damian.

Dream incubation also goes on today in some Hindu temples, notably the temple of Shiva at Tarakeshwar in West Bengal,[42] and also in some Sufi shrines, as I saw for myself at a *dargah* near Hyderabad. Families were sleeping together in little groups in the courtyard surrounding the shrine, accompanying a sick member of the family, with the intention of receiving healing dreams. In the morning, an *imam* helped interpret the dreams they received. I was told that many people were in fact healed as a result of these dreams.

As part of the revival of pilgrimage in Europe (*SSP*, chapter 7), pilgrims are sleeping in ancient churches. Now that sleeping in churches and other sacred places has become feasible, there is a new opportunity for investigating the healing and oracular dreams in holy places. How do the dreams of modern Western people in European holy places compare with those in traditional places of dream incubation in Hindu temples, Sufi shrines and Greek Orthodox churches? Are some holy places more effective than others in inducing healing or oracular dreams? And if so, can the power of the place be distinguished from suggestion and expectation? Conversely, can suggestion and expectation be distinguished from the power of the place?

Gifts and quests

In the spiritual realm, nothing is ours by right. Although spiritual practices can often help us, the results depend on grace. They are not automatic. Deep insights through mystical experiences sometimes arise spontaneously. Everything we receive is a gift. Being thankful helps to maintain the flow of the spirit. The more we receive, the more we can give. And the more we give, the more we can receive.

Finally, spiritual practices may help to illuminate our understanding of ultimate consciousness, *sat-chit-ananda*, God, or whatever name we prefer to give to the ground of all being. Our own being and experience are, according to many traditions, directly linked to the source of All. Our inner consciousness, Atman, is Brahman, as the Hindu tradition puts it. In the Christian tradition, humans are images of God, not because God is like a gigantic human, but because human beings participate in God's Being, as expressed through the creative Logos and the Holy Spirit.

So does all nature. Both humans and the natural world participate in the divine; we are all sustained from moment to moment by divine Being, the ground of all existence, and participate in the Logos and the Spirit, the basis of our lives and minds.

In many religious traditions, we are encouraged to seek God, or search for Truth:

Happy are those . . . who seek him with their whole heart. Psalm 119: 2 [43]

Ask, and it will be given you; search, and you will find; knock, and the door will be opened for you. Luke 11: 9 [44]

God, ever mighty and majestic is he, says, 'I am present in my servant's thought of me, and I am with him when he remembers me.' If he approaches me by a hand's breadth, I draw near to him by an arm's length, and if he draws near to me by an arm's length, I draw near to him by a fathom. If he comes to me walking, I come to him running. Ibn Arabi [45]

For those who see me everywhere and see all things in me, I am never lost, nor are they ever lost to me. Bhagavad Gita 6:30 [46]

If I were asked to define the Hindu creed, I should simply say:
Search after truth through non-violent means. Hinduism is a
relentless pursuit after truth . . . Mahatma Gandhi [47]

This search is now potentially open to everyone, and spiritual practices from many different traditions are accessible as never before. Through a combination of science and spiritual practices, this ancient quest is taking new forms. Spiritual evolution is accelerating.

Notes

Introduction

1 Sheldrake, 2012.
2 Damasio, 1994.
3 Lopez and Snyder (eds), 2009.
4 Ibid.
5 *The Journal for Consciousness Studies*, one of the key publications in the field, was first established in 1994.
6 http://www.openculture.com/2016/08/philosopher-sam-harris-leads-you-through-a-26-minute-guided-meditation.html. Retrieved 10 September 2018.
7 The pioneer of this field of enquiry was William James in his classic book *The Varieties of Religious Experience*, which were the Gifford Lectures in 1901–2 (James, 1960).

Chapter 1: The Spiritual Side of Sports

1 Partridge, 1961, p. 503.
2 *Oxford Shorter English Dictionary*, 1975, p. 1604.
3 Papineau, 2017, p. 235.
4 Ibid., p. 236.
5 Ibid., p. 237.
6 Darwin, 1885, p. 496.
7 Ibid., p. 500.
8 Ibid., p. 562.
9 Block and Dewitte, 2009.
10 Wood and Stanton, 2012.
11 Ibid.
12 Ibid., pp. 149–50.

13 Neave and Wolfson, 2003.

14 Block and Dewitte, 2009.

15 Blanchard, 1995, p. 153.

16 Ibid., p. 1.

17 Ibid., pp. 107–8.

18 Ibid., p. 110.

19 Ibid., p. 167.

20 Plato, *The Laws*, pp. 795–6.

21 Muhammad, 1998, p. 63.

22 Blanchard, 1995, p. 240.

23 Ibid., p. 191.

24 Meggyesy, 2005, pp. 5–6.

25 Ibid., pp. 47–8.

26 Ibid., p. 72.

27 Ibid., p. 73.

28 For example, https://www.youtube.com/watch?v=u65DJiIxWVo&-feature=youtu.be. Retrieved 5 May 2018.

29 Murphy, 2011, p. 67.

30 Murphy and White, 1978, p. 121.

31 Ibid., p. 122.

32 Murphy, 1992.

33 Sportsenergygroup.com. Retrieved 13 June 2017.

34 Csikszentmihalyi, 2002.

35 Csikszentmihalyi et al., 2005, pp. 598–698.

36 Ibid., p. 58.

37 Murphy and White, 1978, pp. 25–6.

38 Jackson and Csikszentmihalyi, 1999, p. 4.

39 Cooper, 1998, p. 34.

40 Jackson and Csikszentmihalyi, 1999, p. 12.

41 Murphy and White, 1978, pp. 30–1.

42 https://www.quora.com/What-is-the-full-context-of-the-quotation-Faster-faster-faster-until-the-thrill-of-speed-overcomes-the-fear-of-death-attributed-to-Hunter-S-Thompson-author-journalist. Retrieved 20 June 2017.

43 https://en.wikipedia.org/wiki/Rider_deaths_in_motorcycle_racing. Retrieved 20 June 2017.

44 https://en.wikipedia.org/wiki/Driver_deaths_in_motorsport. Retrieved 20 June 2017.

45 http://mccabism.blogspot.co.uk/2010/01/ayrton-senna-and-religious-experience.html. Retrieved 5 February 2018.

46 Hawkins, 1983.

47 Murphy and White, 1978, p. 53.

48 Sheldrake, 2013.

49 Novak, 1976, pp. 135–6.

50 Sheldrake, 2011b.

51 Leskowitz, 2014, p. 145.

52 E.g., http://classicallytrained.net/flow-in-video-games/. Retrieved 13 June 2017.

53 https://en.wikipedia.org/wiki/Video_game_addiction. Retrieved 13 June 2017.

54 Meggyesy, 2005, p. 231.

55 Herrigel, 1988.

56 Payne, 1981, p. 32.

57 Ibid., p. 34.

58 Ibid., p. 12.

59 Coffey, 2008, p. 22.

60 Ibid., pp. 91–2.

61 Ibid., p. 25.

62 Ibid., p. 66.

63 Nestor, 2015, p. 31.

64 Ibid., pp. 89–90.

65 Ibid., p. 130.

66 Ibid., p. 29.

67 Ibid., p. 89.

68 Ibid., p. 37.

69 Ibid., p. 88.

70 Ibid., pp. 227–8.

71 Ibid., p. 84.

72 http://bleacherreport.com/articles/417797-the-evolution-of-the-ufc. Retrieved 22 June 2017.

73 E.g., http://www.askaprepper.com/best-mma-self-defense-techniques/. Retrieved 10 September 2018.

74 Leskowitz, 2014.

75 Xu et al., 2004.

76 Huizinga, 1949.

Chapter 2: Learning from Animals

1 http://newspapers.bc.edu/cgi-bin/bostonsh?a=d&d=BOSTONSH 19160701-01.2.26. Retrieved 14 July 2017.

2 http://rwww.rspb.org.uk/about/index.aspx. Retrieved 2 May 2016.

3 Pontzer, 2012.

4 Mithen, 1996.

5 Ehrenreich, 1997.

6 Ibid.

7 Eliade, 1964; Burkert, 1996.

8 May and Marwaha, 2015.

9 Eliade, 1964, p. 94.

10 Masson, 1997.

11 Driscoll and Macdonald, 2010.

12 Morell, 1997.

13 Fiennes and Fiennes, 1968.

14 Serpell, 1983.

15 Ibid.

16 Lindberg et al., 2005.

17 Trut et al., 2004.

18 Galton, 1865.

19 Kerby and Macdonald, 1988.

20 Driscoll et al., 2009.

21 Clutton-Brock, 1981, p. 110.

22 Fiennes and Fiennes, 1968.

23 Source: UK Pet Food Manufacturers' Association.

24 Source American Veterinary Medical Association: https://www.avma.org/KB/Resources/Statistics/Pages/Market-research-statistics-US-pet-ownership.aspx. Retrieved 14 July 2017.

25 Source: Pet Food Manufacturers' Association: http://www.pfma.org.uk/pet-population-2015. Retrieved 14 July 2017.

26 Source: Euromonitor International: http://blog.euromonitor.com/2015/07/german-pet-population-it-is-bigger-than-you-thought.html. Retrieved 14 July 2017.

27 https://www.thestar.com/news/insight/2017/07/08/do-therapy-animals-actually-help-reduce-stress-researchers-are-conflicted.html. Retrieved 19 February 2018.

28 For example, Summerfield, 1996.

29 For example, Phear, 1997.

30 Darwin, 1859, p. 179.

31 Masson, 1997, p. 86.

32 Hart, 1995; Rennie, 1997.

33 For the most influential statement of this point of view, see Dawkins, 1976.

34 The most systematic exposition of this theory is that of Wilson, 1980.

35 For a discussion of the extent to which giving alarm signals can be dangerous for the individual though beneficial to the group, see Ridley, 1996.

36 Serpell, 1991.

37 *NIH News in Health*, February 2009.

38 Karsh and Turner, 1988.

39 Ibid.

40 Hart, 1995; Dossey, 1997b.

41 Lynch and McCarthy, 1969.

42 Friedmann, 1995.

43 Hart, 1995.

44 Ibid.

45 Hart, 1995.

46 Dossey, 1997b.

47 Ormerod, 1996.

48 For example, Paul and Serpell, 1996.

49 Howick, 2017.

50 Kiecolt-Glaser and Glaser, 2002.

51 Humphrey, 2002, chapter 19.

52 Sheldrake, 2012, chapter 9.

53 Richard Dawkins, Daniel Dennett and many other militant materialists are fellows of the Committee for Skeptical Inquiry, an advocacy organisation that opposes 'claims of the paranormal': https://www.csicop.org/about/csi_fellows_and_staff. Retrieved 11 September 2018.

54 May and Marwaha, 2015.

55 Sheldrake and Smart, 1998, 2000a, 2000b.

56 Sheldrake, 2011b.

57 Ibid.

58 Bender, 2014, pp. 44–5.

59 Ibid., p. 45.

60 Sheldrake, 2013.

61 Ibid.

62 This research is summarised in Sheldrake, 2013.

63 Sheldrake and Smart, 2003.

64 Sheldrake and Smart, 2005; Sheldrake et al., 2009.

65 Sheldrake, 2013.

66 Sheldrake, 1990.

67 For example, Nagel, 2012.

68 Jayaram V., 'The sacred animals of Hinduism', https://www.hindu-website.com/hinduism/essays/sacred-animals-of-hinduism.asp. Retrieved 12 September 2018.

69 Aquinas, 2009, p. 83.

Chapter 3: Fasting

1 MacCulloch, 1912.

2 Ibid.

3 Ibid., p. 762.

4 Ibid.

5 Quoted in Muhammed, 2017, p. 29.

6 Beresford, 1987.

7 Anwyl, 1909.

8 Beresford, 1987.

9 Sinclair, 1911.

10 Longo and Mattson, 2014.

11 Ibid.

12 Ibid.

13 Kerndt et al., 1982.

14 Stewart and Fleming, 1973.

15 Fung, 2016, p. 77.

16 Ibid., p. 51.

17 Ibid., p. 52.

18 http://www.who.int/mediacentre/factsheets/fs311/en/. Retrieved 19 April 2017.

19 http://www.who.int/mediacentre/factsheets/fs312/en/. Retrieved 19 April 2017.

20 Bostock et al., 2017.

21 https://www.diabetes.co.uk/keto/ketogenic-diet-and-mental-health.html. Retrieved 12 September 2018.

22 Brown, 2007.

23 Kamal et al., 2016.

24 https://www.drugs.com/illicit/ghb.html. Retrieved 12 September 2018.

25 Fung, 2016.

26 Jeremiah, 2010.

27 http://indianexpress.com/article/explained/the-jain-religion-and-the-right-to-die-by-santhara/. Retrieved 19 April 2017.

28 https://www.psychologytoday.com/blog/changing-the-way-we-die/201407/fasting-death. Retrieved 20 April 2017.

29 Ibid.

30 Harlan, 1991.

31 Dasgupta, 2010.

32 Thurston, 1952.

33 Ibid., p. 377.

34 Ibid., p. 366.

35 Ibid, p. 384.

36 Fung, 2016.

Chapter 4: Cannabis, Psychedelics and Spiritual Openings

1 Graber, C., 2008. 'Fact or fiction? Animals like to get drunk', https://www.scientificamerican.com/article/animals-like-to-get-drunk/. Retrieved 10 September 2017.

2 Carrigan, M., 2014. 'Human taste for alcohol linked to apes eating rotten fruit', http://www.cbc.ca/news/technology/human-taste-for-alcohol-linked-to-apes-eating-rotten-fruit-1.2871052. Retrieved 12 September 2017.

3 Buhner, 1998, p. 24.

4 Ibid., p. 4.

5 http://www.medicaldaily.com/psychedelic-drug-use-united-states-common-now-1960s-generation-245218. Retrieved 12 September 2017.

6 Hogan, E., 'Turn on, tune in, drop by the office', *The Economist* August/September 2017, https://www.1843magazine.com/features/turn-on-tune-in-drop-by-the-office; see also Kuchler, H., 2017. 'How Silicon Valley Rediscovered LSD', *Financial Times*, 10 August 2017, https://www.ft.com/content/0a5a4404-7c8e-11e7-ab01-a13271d1ee9c. Retrieved 12 September 2017.

7 Summarised in Pollan, 2018.

8 https://www.ancient-origins.net/myths-legends/strange-life-al-khidr-legendary-immortal-prophet-mystic-trickster-and-sea-spirit-020673. Retrieved 17 September 2018.

9 http://khidr.org/cannabis.htm. Retrieved 12 September 2017.

10 Akhtar, A., 'Pakistan's "heretical" Muslims', *Guardian*, 23 October 2009, https://www.theguardian.com/commentisfree/belief/2009/oct/23/pakistan-sufis-terrorism-shrines. Retrieved 12 September 2017.

11 https://en.wikipedia.org/wiki/Rastafari. Retrieved 12 September 2017.

12 https://pointsadhsblog.wordpress.com/2015/06/11/the-use-of-marijuana-in-the-rastafari-religion/. Retrieved 14 September 2017.

13 Crawford, 2004.

14 Quickfall and Crockford, 2006.

15 McPartland et al., 2005.

16 Pagotto et al., 2006.

17 Fernández-Ruiz et al., 2013.

18 Owens, 2015.

19 Boserman, 2009.

20 Englund et al., 2013.

21 http://www.christianitytoday.com/pastors/2013/november-online-only/what-would-jesus-smoke.html. Retrieved 16 September 2017.

22 http://abcnews.go.com/Health/rabbi-ties-jewish-faith-medical-marijuana-dispensary/story?id=20348883. Retrieved 16 September 2017.

23 http://hightimes.com/culture/onward-christian-stoners/. Retrieved 16 September 2017.

24 http://hightimes.com/culture/music/mazel-tov-jews-get-chai-at-las-first-cannabis-havdalah-party/. Retrieved 16 September 2017.

25 Ibid., p. xxi.

26 Shulgin and Shulgin, 1991.

27 Ibid., p. xx.

28 Ibid., p. 736.

29 Saunders, 1995, p. 14.

30 https://en.wikipedia.org/wiki/Albert_Hofmann#cite_note-10.

31 Ott, 1993, p. 127.

32 Ibid., pp. 141–4.

33 Shulgin and Shulgin, 1997.

34 Ibid., p. 405.

35 Luke, 2017, chapter 3.

36 Rios and Janiger, 2003.

37 Huxley, 1994, p. 13.

38 Ibid., pp. 70–1

39 Ibid., p. 72.

40 Ibid., pp. 73–4.

41 New Revised Standard Version.

42 Huxley, 1994, p. 76.

43 Quoted in Rios and Janiger, 2003, p. 134.

44 Quoted in Sheldrake et al., 2005, p. 23.

45 Ibid., pp. 23–4.

46 Luke, 2017, p. 224; Carhart-Harris et al., 2012; Speth et al., 2016.

47 Carhart-Harris et al., 2016.

48 Roseman et al., 2016.

49 Ibid.

50 Kaelen et al., 2015.

51 Ly et al., 2018.

52 Huxley, 1994, p. 8.

53 Rios and Janiger, 2003, p. 10.

54 Sessa, 2017, p. 29.

55 Rios and Janiger, 2003, p. 69.

56 Ibid., p. 70.

57 Carter, 2010.

58 D'Alviella, 1914, pp. 317–18.

59 Ibid., p. 318.

60 Ibid., p. 317.

61 Ibid., p. 317.

62 Grof and Grof, 1980, p. 25.

63 Heyes, 2000, p. 14.

64 Ibid., p. 15.

65 Unlike the other psychedelics discussed in this chapter, ketamine primarily works through the binding sites for the neurotransmitter glutamate.

66 Sessa, 2017, p. 62.

67 Jansen, 1996.

68 Strassman, 2001.

69 Ibid., p. 274.

70 Ibid., p. 277.

71 Ibid., pp. 233–4.

72 Ibid., p. 306.

73 Quoted in Shanon, 2008.

74 Shanon, 2005.

75 https://www.theguardian.com/travel/2016/jun/07/peru-ayahuasca-drink-boom-amazon-spirituality-healing. Retrieved 5 October 2017.

76 https://www.newyorker.com/magazine/2016/09/12/the-ayahuasca-boom-in-the-u-s. Retrieved 5 October 2017.

77 For example, http://www.ayahuasquero.com/structure.html. Retrieved 5 October 2017.

78 http://afamiliajuramidam.org/english/mestre_irineu_english.htm. Retrieved 21 June 2018.

79 http://www.maps.org/news-letters/v03n4/03429aya.html. Retrieved 19 September 2018.

80 http://www.religionnewsblog.com/category/uniao-do-vegetal. Retrieved 5 October 2017.

81 Shanon, 2008.

82 Strassman, 2014.

83 For example, McKenna, 1993.

84 Hancock, 2005.

85 Strassman, 2001, p. 199.

86 For example, https://www.quora.com/What-does-it-mean-to-see-Ganesha-in-your-dream. Retrieved 6 June 2018.

87 Sheldrake, Abraham and McKenna, 2005, p. 33.

88 For a discussion see Sheldrake, 2013, chapters 15 and 16.

89 Holecek, 2016; Norbu, 2012.

90 Ehrenreich, 1997.

91 The Bible: Book of Daniel, chapter 2.

92 Sheldrake, 2009.

93 Naranjo, 2006.

94 Hofmann, 1983.

95 Fábregas, 2010.

96 Bouso, 2012.

97 E.g., Zaehner, 1957.

Chapter 5: Powers of Prayer

1 New Revised Standard Version.

2 Tearfund Prayer Survey by ComRes, published 14 January 2018, table 15, http://www.comresglobal.com/polls/tearfund-prayer-survey/. Retrieved 15 January 2018.

3 African Religions: http://www.deathreference.com/A-Bi/African-Religions.html. Retrieved 6 April 2018.

4 Fallaize, 1918, p. 154.

5 Boyer, 2002, p. 156.

6 D'Arcy, 1918, p. 171.

7 In Islam, Judaism and in Orthodox churches, 'Amen' is pronounced *ameen*. For a discussion of the effects of these two forms, see Sheldrake, 2017, chapter 6.

8 https://catholiconline.shopping/products/slow-burning-virtual-candle. Retrieved 23 October 2017.

9 Prayers to the Angels of God: http://www.catholic.org/saints/angels/angelprayer.php.

10 Frazer, 1918.

11 Luhrmann, 2012, pp. 55–6.

12 Ibid., p. 296.

13 See Sheldrake, 2012.

14 Lorimer, 2017.

15 Murphy, 2015, pp. 375–6.

16 Tearfund Prayer Survey by ComRes, published 14 January 2018, table 15, http://www.comresglobal.com/polls/tearfund-prayer-survey/. Retrieved 15 January 2018.

17 Boyer, 2002, p. 179.

18 Ibid., p. 179.

19 New Revised Standard Version.

20 Taylor, 2007.

21 Foster, 2010, pp. 189–90.

22 Brown, 2012, pp. 1–2.

23 Ibid., p. 14.

24 Ibid., p. 25.

25 Ibid., p. 279.

26 Marchant, 2016.

27 Brown, 2012, p. 281.

28 Benson et al., 2006.

29 Schwartz and Dossey, 2010.

30 Ibid.

31 Ibid., p. 15.

32 Howick, 2017.

33 Marchant, 2016, p. 286.

34 Lee et al., 2017.

35 Edwards, 2016.

36 Dossey, 1993, p. 205.

37 Koenig et al., 2012, pp. 58, 62.

38 Ibid., p. 60.

39 Ibid., p. 61.

40 Ibid., pp. 61–2.

41 Ibid., pp. 601–2.

42 Ibid., chapter 9.

43 Ibid., p. 143.

44 Horowitz, 2014.

45 Allen, 1917, p. 360.

46 Horowitz, 2014, p. 199.

47 Peale, 1955, p. 9.

48 Ibid., pp. 11, 13.

49 Horowitz, 2014.

50 Seligmen, 2002, p. 98.

51 Ibid., p. 288.

52 Horowitz, 2014.

53 Dossey, 1997, p. 62.

54 Ibid., pp. 68–9.

55 Partridge, 1961, p. 50.

56 New Revised Standard Version.

57 O'Donohue, 2007, pp. 15–17.

58 Ibid., p. 37.

59 If you would like some well-prepared guidelines for each day, see http://www.trypraying.co.uk. Retrieved 16 November 2017.

Chapter 6: Holy Days and Festivals

1 Cook, 2017.
2 Hancock, 2005.
3 Vitebsky, 2005, pp. 11–12.
4 James, 1961.
5 Ibid., p. 36.
6 Webster, 1918, p. 889.
7 Ibid., p. 890.
8 Quoted in Abrahams, 1918, p. 891.
9 New Revised Standard Version.
10 New Revised Standard Version.
11 Heschel, 1951, p. 10.
12 Ibid., pp. 90–1.
13 https://en.wikipedia.org/wiki/Sunday_Trading_Act_1994. Retrieved 5 December 2017.
14 Allender, 2009, p. 50.
15 https://www.dur.ac.uk/news/newsitem/?itemno=28980. Retrieved 5 December 2017.
16 Jabr, F., 'Why your brain needs more downtime', *Scientific American*, 15 October 2013, https://www.scientificamerican.com/article/mental-downtime/. Retrieved 5 December 2017.
17 http://edition.cnn.com/2013/01/11/health/sleeth-take-day-off/index.html. Retrieved 5 December 2017.
18 https://www.psychologytoday.com/blog/the-power-rest/201705/rest-success. Retrieved 5 December 2017.
19 James, 1961, pp. 175–6.
20 Ibid., p. 229.
21 Hutton, 1996.
22 Ibid., p. 3.
23 Ibid., p. 121.
24 Ibid.
25 Quoted Hopkins, 1912, p. 869.
26 Heschel, 1951, pp. 7–8.
27 Rundle Clark, 1959.

28 Frazer, 1917, vol. I, p. 6.

29 Ibid., part V, p. 40.

30 Ibid., 1920, part III, p. 9.

31 Hutton, 1996, p. 226.

32 Ibid., p. 228.

33 Ibid., p. 229.

34 Whistler, 1947, p. 143.

35 Hutton, 1996, chapters 23 and 24.

36 Ibid., p. 220.

37 Whistler, 1947, p. 141.

38 Roud, 2006 , p. 199.

39 Crawley, 1915, p. 503.

40 Roud, 2006, p. 301.

41 Whistler, 1947, p. 169.

42 Hutton, 1996, p. 313.

43 Coomaraswamy, 1935.

44 Barns, 1915, p. 621.

45 http://news.gallup.com/poll/193271/americans-believe-god.aspx. Retrieved 23 December 2017.

46 http://www.brin.ac.uk/wp-content/uploads/2017/01/No-16-January-2017.pdf. Retrieved 23 December 2017.

47 Matthew Fox, personal communication, 14 April 2014.

48 Quoted in Fox and Sheldrake, 1996, p. 78.

49 James, 1961, p. 227.

50 Hutton, 1996, chapter 35.

51 Ibid., p. 371.

52 Ibid., chapter 36.

53 Ibid., p. 383.

54 Steger, 2009.

55 For a detailed discussion of this hypothesis, see Sheldrake, 2011a.

56 See for example, http://www.sabbathmanifesto.org. Retrieved 28 June 2018.

Chapter 7: Cultivating Good Habits, Avoiding Bad Habits, and Being Kind

1 Peterson and Park, 2009.

2 Peterson and Seligman, 2004, chapter 2.

3 Botton, 2012, p. 87.

4 Ibid., p. 88.

5 Ibid., p. 33.

6 Ibid., p. 29.

7 Phillips and Taylor, 2010, p. 25.

8 For a discussion of this prevalent attitude, see Wiseman, 2018.

9 Phillips and Taylor, 2009, p. 25.

10 Williams, G., 1988, quoted in Waal, 2009, p. 9.

11 Dawkins, 1976, p. 21.

12 Batson et al., 2009, pp. 417–26.

13 Waal, 2009, p. 5.

14 Turnbull, 2015, pp. 104–6.

15 Costa, 2006, p. 24.

16 Hölldobler and Wilson, 2009, pp. 427–30.

17 Pérez-Manrique and Gomila, 2018, p. 265.

18 Darwin, 1885, p. 98.

19 Ibid.

20 Duque, et al., 2018.

21 Ibid., p. 257.

22 Ibid., p. 15.

23 Ibid., p. 19.

24 Pérez-Manrique and Gomila, 2018.

25 https://www.youtube.com/watch?v=Yy-3DWRdLh8. Retrieved 27 March 2018.

26 https://www.youtube.com/watch?v=9b29X3zOIDc. Retrieved 27 March 2018.

27 Hellinger, Weber and Beaumont, 1998, p. 5.

28 Ibid., p. 10.

29 Ibid., pp. 10–11.

30 Moral virtues and vices. https://www.al-islam.org/jami-al-saadat-

the-collector-of-felicities-muhammad-mahdi-ibn-abi-dharr-al-naraqi/
moral-virtues-vices. Retrieved 22 November 2018.

31 Fox and Sheldrake, 1996, p. 125.

32 Fox, 1999, p. 316.

33 Alexander, 1920.

34 Fox, 1999, p. 240.

35 Ibid., p. 265.

36 Ibid., pp. 167–8.

37 12-step success rates. https://luxury.rehabs.com/12-step-programs/
success-rates/. Retrieved 10 April 2018.

38 Brand, 2017, p. 135.

39 Ibid., p. 24.

40 Ibid., p. 25.

41 New Revised Standard Version.

42 Cohn, 1970.

43 Gray, 2018, p.73.

44 See the discussion in Sheldrake, 1990, pp. 29 –32.

45 Whitehouse, 2018.

46 Ibid., p. 421.

47 Surahs 2:143; 9:128.

48 Sahih Muslim 2593

49 Quoted in Phillips and Taylor, 2010. p. 41.

50 Ibid., pp. 12–13.

51 Ibid., p. 41.

52 Boehm and Lyubomirsky, 2009, p. 672.

53 Otake et al., 2006.

54 Random acts of kindness. https://www.randomactsofkindness.org/
the-science-of-kindness. Retrieved 11 April 2018.

55 Tomasello, 2009.

56 Warneken and Tomasello, 2007.

57 Rilling et al., 2002.

58 Dalai Lama, 2001, p. 23

59 Ibid., pp. 53–4

60 Leblanc, 2010.

Chapter 8: Why Do Spiritual Practices Work?

1 Hay, 2006.

2 Hill, 2014. 'Finding God in a seizure: The link between temporal lobe epilepsy and mysticism', http://www.abc.net.au/radionational/programs/archived/encounter/the-link-between-temporal-lobe-epilepsy-and-mysticism/5956982. Retrieved 2 May 2018.

3 Harris, 2014.

4 Blackmore, 2011.

5 Norbu, 2013.

6 Mikulas, 2007.

7 Batchelor, 2011.

8 *Dhammapada*, verse 203. 'Hunger is the greatest ailment, khandhas are the greatest ill. The wise, knowing them as they really are, realise Nibbana, the greatest bliss.'

9 Mallinson and Singleton, 2017, p. xiii.

10 Mascaro (trans), 1965, p. 51.

11 Griffiths, 1982, p. 27.

12 Ibid., p. 190.

13 Hart, 2013, pp. 258–259

14 For a historical summary of field concepts in biology, see Sheldrake, 2009.

15 Popper and Eccles, 1977, pp. 5–7.

16 Jenner, 2014.

17 https://www.swpc.noaa.gov/products/space-weather-advisory-outlook. Retrieved 30 June 2018.

18 Hawking, 1988, p. 174.

19 Krauss, 2012, p. 153.

20 Wroe, 2007.

21 http://www.philosophyforlife.org/the-spiritual-experiences-survey/. Retrieved 9 November 2016.

22 'All his materials', aeon. https://aeon.co/essays/a-rare-interview-with-philip-pullman-the-religious-atheist. Retrieved 6 June 2018.

23 Pollan, 2018, p. 413.

24 E.g., Nagel, 2012.

25 Sheldrake, 1990.

26 Kretzmann, 'Philosophy of mind', In: Kretzmann, N. and Stump, E. (eds), 1993.

27 Quoted by and discussed by Pope Francis, 2015.

28 Milne, 2013, p. 67.

29 Sheldrake, 1990.

30 Sheldrake, 2011a.

31 Fox, 1988, p. 57.

32 Ibid., p. 57.

33 Sheldrake, Abraham and McKenna, 2005, pp. 32–3.

34 Sheldrake, 2011a, chapter 14.

35 McFarland (ed.), 1981, p. 141.

36 E.g., https://mindworks.org/meditation-knowledge/how-many-people-meditate/. Retrieved 14 June 2018.

37 Pew Forum, 2011. 'Global Christianity: A Report on the Size and Distribution of the World's Christian Population', http://www.pewforum.org/2011/12/19/global-christianity-exec/ Retrieved 22 November 2018.

38 Trubshaw, 2017.

39 New Revised Standard Version.

40 Authorised Version (King James Bible).

41 Trubshaw, 2017.

42 http://kellybulkeley.org/dreams-and-healing-in-west-bengal/. Retrieved November 22 2018.

43 New Revised Standard Version.

44 New Revised Standard Version.

45 Ibn 'Arabī, 2004, p. 24.

46 https://www.holy-bhagavad-gita.org. Retrieved June [??]

47 Quoted at https://www.goodreads.com/quotes/138933-if-i-were-asked-to-define-the-hindu-creed-i. Retrieved 12 June 2018.

Acknowledgements

This book, like my previous book, *Science and Spiritual Practices*, has been much influenced by my wife, Jill Purce, who has been teaching spiritual practices for many years, especially chanting, singing and healing ceremonies, and who has shown over and over again that such practices are potentially open to all, whether they think of themselves as religious or not. She also suggested the title for this book. I am very grateful to her, and also to our sons Merlin and Cosmo, for their encouragement and support in this project.

I have been thinking about many of the themes in this book for a long time, and have been helped by many conversations and exchanges of correspondence. I am especially grateful to Ralph Abraham, David Abram, Marc Andrus, Linda Bender, Bernard Carr, Angelika Cawdor, Deepak Chopra, John Cobb, Larry Dossey, Lindy Dufferin and Ava, Amanda Feilding, Peter Fenwick, Addison Fischer, Claire Foster-Gilbert, Matthew Fox, Rob Freeman, Peter Fry, Adele Getty, David Ray Griffin, the late Father Bede Griffiths, Stephan Harding, Guy Hayward, Bert Hellinger, Liz Hosken, Nicholas Humphrey, the late Francis Huxley, Robert Jackson, Satish Kumar, Natuschka Lee, James Le Fanu, David Lorimer, David Luke, Luis Eduardo Luna, Nancy Lunney, the late Terence McKenna, Brother Martin Kuvarupu, Moez Masoud, Ralph Metzner, John and Alison Milbank, the late Michael Moore, Michael Murphy, Richard Perl, Edward Posey, Dean Radin, Anthony Ramsay, Edward St Aubyn, Marilyn Schlitz, Brother David Steindl-Rast, Rick Tarnas, Mark Vernon, Cassandra Vietan, Ian and Victoria Watson, Fraser Watts, Andrew Weil, Jason Wentworth, Michael Williams and Gordon Wheeler.

This book is dedicated with much gratitude to Peggy Taylor and her husband Rick Ingrasci, who have helped and inspired me – and all my family – for more than thirty-five years.

I thank the staff and students at the centres where I have taught on many occasions, and where I have learned so much, especially the Esalen Institute, in Big Sur, California; the Institute of Noetic Sciences, in Petaluma, California; Hollyhock, on Cortes Island, British Columbia; and Schumacher College, in Dartington, Devon.

I am grateful for the financial support for the research that helped me write this book, from Addison Fischer and the Planet Heritage Foundation, of Naples, Florida and the Gaia Foundation, London; from the Watson Family Foundation and the Institute of Noetic Sciences; and from the Peter Hesse Foundation in Düsseldorf, Germany.

Many thanks to Pam Smart, my research assistant, who has worked with me for more than twenty-four years and has helped me in many ways. I am also grateful to Guy Hayward for all his help in this project, made possible by a postdoctoral research fellowship funded through the Scientific and Medical Network, and to Sebastian Penraeth, my webmaster.

I am grateful to my editor, Mark Booth, at Hodder & Stoughton, in London, for his encouragement to write this book, and for his thoughtful editing.

I thank all those who have commented on drafts of this book, especially Guy Hayward, Douglas Hedley, Cosmo and Merlin Sheldrake, Mark Vernon and Jason Wentworth.

Bibliography

Abrahams, I., 'Sabbath (Jewish)', In: Hastings, J. (ed.), *Encyclopaedia of Religion and Ethics*, Vol. 10, Clark, Edinburgh, 1918.

Alexander, A.B.D., 'Seven deadly sins', In: Hastings, J. (ed.), *Encyclopaedia of Religion and Ethics*, Vol. 11, Clark, Edinburgh, 1920.

Allen, A.L., 'New Thought', In: Hastings, J. (ed.), *Encyclopaedia of Religion and Ethics*, Vol. 9, Clark, Edinburgh, 1917.

Allender, D.B., *Sabbath*, Thomas Nelson, Nashville, TN, 2009.

Alviella, G. d', 'Initiation (Introductory and Primitive)', In: Hastings, J. (ed.), *Encyclopaedia of Religion and Ethics*, Vol. 7, Clark, Edinburgh, 1914.

Anwyl, E., 'Asceticism (Celtic)', In: Hastings, J. (ed.), *Encyclopaedia of Religion and Ethics*, Vol. 2, Clark, Edinburgh, 1909.

Aquinas, T., *Compendium of Theology*, Oxford University Press, Oxford, 2009.

Barns, T., 'Michaelmas', In: Hastings, J. (ed.), *Encyclopaedia of Religion and Ethics*, Vol. 9, Clark, Edinburgh, 1915.

Batchelor, S., *Confession of a Buddhist Atheist*, Spiegel and Grau, London, 2011.

Batson, C. D., Ahmad, N. and Lishner, D.A., 'Empathy and altruism', In: Lopez, S.J. and Snyder, C.R. (eds), *The Oxford Handbook of Positive Psychology* (2nd ed.), Oxford University Press, New York, NY, 2009.

Bender, L., *Animal Wisdom: Learning from the Spiritual Lives of Animals*, North Atlantic Books, Berkeley, CA, 2014.

Benson, H., Dusek, J.A. et al., 'Study of the therapeutic effects of intercessory prayer (STEP) in cardiac bypass patients: a multicenter randomized trial of uncertainty and certainty of receiving intercessory prayer', *American Heart Journal*, 2006, 151, 934–42.

Beresford, D., 'Ten Men Dead: The Story of the 1981 Irish Hunger Strike', *Atlantic Monthly*, New York, 1987.

Blackmore, S., *Zen and the Art of Consciousness*, Oneworld Publications, London, 2011.

Blanchard, K., *The Anthropology of Sport: An Introduction*, Bergin and Garvey, Westport, CT, 1995.

Block, A. de and Dewitte, S., 'Darwinism and the cultural evolution of sports', *Perspectives in Biology and Medicine*, 2009, 52, 1–16.

Boehm, J.K. and Lyubomirsky, S., 'The promise of sustainable happiness', In: Lopez, S.J. and Snyder, C.R. (eds), *The Oxford Handbook of Positive Psychology* (2nd ed.), Oxford University Press, New York, NY, 2009.

Boserman, C., 'Diaries from cannabis users: an interpretative phenomenological analysis', *Health*, 2009, 13, 429–48.

Bostock, E.C.S., Kirkby, K.C. and Taylor, B.V.M., 'The current status of the ketogenic diet in psychiatry', *Frontiers in Psychiatry*, 20 March 2017, doi: 10.3389/fpsyt.2017.00043.

Botton, A. de, *Religion for Atheists: A Non-believer's Guide to the Uses of Religion*, Hamish Hamilton, London, 2012.

Bouso, J.C. et al., 'Personality, psychopathology, life attitudes and neuropsychological performance among ritual users of ayahuasca: a longitudinal study', *PLoS ONE*, 2012, 7, 1–13: http://dx.doi.org/10.1371/journal.pone.0042421.

Boyer, P., *Religion Explained: The Human Instincts that Fashion Gods, Spirits and Ancestors*, Vintage, London, 2002.

Brand, R., *Recovery: Freedom from Our Addictions*, Bluebird Books, London, 2017.

Brown, A. J., 'Low-carb diets, fasting and euphoria: is there a link between ketosis and gamma-hydroxybutyrate (GHB)?', *Medical Hypotheses*, 2007, 68, 268–71.

Brown, C.G., *Testing Prayer: Science and Healing*, Harvard University Press, Cambridge, MA, 2012.

Buhner, S.H., *Sacred and Herbal Healing Beers: The Secrets of Ancient Fermentation*, Brewers Publications, Boulder, CO, 1998.

Burkert, W., *Creation of the Sacred: Tracks of Biology in Early Religions*, Harvard University Press, Cambridge, MA, 1996.

Carhart-Harris, R.L. et al., 'Neural correlates of the psychedelic state as determined by fMRI studies with psilocybin', *Proceedings of the National Academy of Sciences (US)*, 2012, 109, 2138–43.

Carhart-Harris, R.L., Muthukumaraswamy, S., Roseman, L. et al., 'Neural correlates of the LSD experience revealed by multi-modal neuroimaging', *Proceedings of the National Academy of Sciences (US)*, 2016, 113, 4853–8.

Carter, C., *Science and the Near-Death Experience: How Consciousness Survives Death*, Inner Traditions, Rochester, NY, 2010.

Chögyal Namkhai Norbu, *Dream Yoga and the Practice of Natural Light*, Snow Lion Publications, Ithaca, NY, 2012.

Chögyal Namkhai Norbu, *Rainbow Body: The Life and Realization of a Tibetan Yogin, Togden Ugyen Tendzin,* North Atlantic Books, Berkeley, CA, 2013.

Clutton-Brock, J., *Domesticated Animals from Early Times*, Heinemann, London, 1981.

Coffey, M., *Explorers of the Infinite*, Tarcher, New York, NY, 2008.

Cohn, N., *The Pursuit of the Millennium: Revolutionary Millenarians and Mystical Anarchists of the Middle Ages*, Granada, London, 1970.

Cook, J., 'Believing and belonging', *The British Museum Magazine*, 2017, 89, 24–8.

Coomaraswamy, A., 'Angel and titan: an essay in Vedic ontology', *Journal of the American Oriental Society*, 1935, 55, 373–418.

Cooper, A., *Playing in the Zone: Exploring the Spiritual Dimensions of Sports*, Shambhala, Boston, MA, 1998.

Costa, J.T., *The Other Insect Societies*, Harvard University Press, Cambridge, MA, 2006.

Crawford, V., 'A homelie herb: medicinal cannabis in early England', *European Journal of Herbal Medicine*, 2004, 6, 5–11.

Crawley, A.E., 'May, midsummer', In: Hastings. J. (ed.), *Encyclopaedia of Religion and Ethics*, Vol. 5, Clark, Edinburgh, 1915.

Csikszentmihalyi, M., *Flow: The Classic Work on How to Achieve Happiness*, Rider, London, 2002.

Csikszentmihalyi, M., Abuhamdeh, S. and Nakamura, J., 'Flow', In: Elliot, A. (ed.), *Handbook of Competence and Motivation*, The Guilford Press, New York, NY, 2005.

D'Arcy, C.F., 'Prayer (Christian)', In: Hastings, J. (ed.), *Encyclopaedia of Religion and Ethics*, Vol. 10, Clark, Edinburgh, 1918.

Dalai Lama, *The Dalai Lama's Book of Love and Compassion*, Thorsons, London, 2001.

Damasio, A., *Descartes' Error: Emotion, Reason and the Human Brain*, Putnam, New York, 1994.

Darwin, C., *On the Origin of Species by Means of Natural Selection*, Murray, London, 1859.

Darwin, C., *The Descent of Man, and Selection in Relation to Sex* (2nd ed.), Murray, London, 1885.

Dasgupta, M., 'DIPAS concludes observational study on "Mataji"', *The Hindu*, 10 May 2010.

Dawkins, R., *The Selfish Gene*, Oxford University Press, Oxford, 1976.

Descola, P., *Beyond Nature and Culture*, University of Chicago Press, 2013.

Dossey, L., *Healing Words: The Power of Prayer and the Practice of Medicine*, HarperSanFrancisco, San Francisco, CA, 1993.

Dossey, L., *Be Careful What You Pray For . . . You Just Might Get It*, HarperSanFrancisco, San Francisco, CA, 1997a.

Dossey, L., 'The healing power of pets: a look at animal-assisted therapy', *Alternative Therapies*, 1997b, 3, 8–15.

Driscoll, C.A. and Macdonald, D.W., 'Top dogs: wolf domestication and wealth', *Journal of Biology*, 2010, 9, 10.

Driscoll, C.A., Macdonald, D.W. and O'Brien, S.J., 'From wild animals to domestic pets, an evolutionary view of domestication', *Proceedings* of the National Academy of Sciences (US), 106 (Supplement 1), 2009, 9971–8.

Duque, J.F., Leichner, W., Ahmann, H. and Stevens, J.R., 'Mesotocin influences pinyon jay prosociality', *Biology Letters*, 14, issue 4, doi: 10.1098/rsbl.2018.0105, 11 April 2018.

Edwards, S., *Healing in a Hospital: Scientific Evidence that Spiritual Healing Improves Health*, Edwards, Amazon, UK, 2016.

Ehrenreich, B., *Blood Rites*, Metropolitan Books, New York, 1997.

Eliade, M., *Shamanism: Archaic Traditions of Ecstasy*, Princeton University Press, Princeton, 1964.

Englund, A. et al., 'Cannabidiol inhibits THC-elicited paranoid symptoms and hippocampal-dependent memory impairment', *Journal of Psychopharmacology*, 2013, 27, 19–27.

Fábregas, J.M. et al., 'Assessment of addiction severity among ritual users of ayahuasca', *Drug and Alcohol Dependence*, 2010, 111, 257–61.

Fallaize, E.N., 'Prayer (Introductory and Primitive)', In: Hastings, J. (ed.), *Encyclopaedia of Religion and Ethics*, Vol. 10, Clark, Edinburgh, 1918.

Fernández-Ruiz, J. et al., 'Cannabidiol for neurodegenerative disorders: important new clinical applications for this phytocannabinoid?', *British Journal of Clinical Pharmacology*, 2013, 75, 323–33.

Fiennes, R. and Fiennes, A., *The Natural History of the Dog*, Weidenfeld and Nicolson, London, 1968.

Foster, C., *Wired for God? The Biology of Spiritual Experience*, Hodder, London, 2010.

Fox, M., *The Coming of the Cosmic Christ*, Harper and Row, San Francisco, CA, 1988.

Fox, M., *Sins of the Spirit, Blessing of the Flesh: Lessons for Transforming Evil in Soul and Society*, Harmony Books, New York, NY 1999.

Fox, M. and Sheldrake, R., *The Physics of Angels: Exploring the Realm Where Science and Spirit Meet*, HarperCollins, San Francisco, CA, 1996.

Frazer, J.G., *The Golden Bough* (3rd ed.), Macmillan, London, 1917–20.

Frazer, J.G., *Folk-Lore in the Old Testament*, Macmillan, London, 1918.

Friedmann, E., 'The role of pets in enhancing human well-being: physiological effects', In: Robinson, I. (ed.), *The Waltham Book of Human-Animal Interaction: Benefits and Responsibilities of Pet Ownership*, Pergamon Press, Oxford, 1995.

Fung, J., *The Complete Guide to Fasting: Heal Your Body Through Intermittent, Alternate-Day and Extended Fasting*, Victory Belt Publishing, Las Vegas, 2016.

Galton, F., 'The first steps towards the domestication of animals', *Transactions of the Ethnological Society of London, New Series*, 1865, 3, 122–38.

Gray, J., *Seven Types of Atheism*, Allen Lane, London, 2018.

Griffiths, B., *The Marriage of East and West*, Collins, London, 1982.

Grof, S. and Grof, C., *Beyond Death: The Gates of Consciousness*, Thames and Hudson, London, 1980.

Hancock, G., *Supernatural: Meetings with the Ancient Teachers of Mankind*, Arrow, London, 2005.

Harlan, L., *Religion and Rajput Women*, University of California Press, Berkeley, CA, 1991.

Harris, S., *Waking Up: Searching for Spirituality Without Religion*, Transworld Publishers, London, 2014.

Hart, D.B., *The Experience of God: Being, Consciousness Bliss*, Yale University Press, New Haven, 2013.

Hart, L.A., 'Dogs as human companions: a review of the relationship', In: Serpell, J. (ed.), *The Domestic Dog*, Cambridge University Press, Cambridge, 1995.

Hawking, S., *A Brief History of Time*, Bantam, London, 1988.

Hay, D., *Something There: The Biology of the Human Spirit*, Darton, Longman and Todd, London, 2006.

Hellinger, B., Weber, G. and Beaumont, H., *Love's Hidden Symmetry: What Makes Love Work in Relationships*, Zeig, Tucker and Theisen, Phoenix, AZ, 1998.

Herrigel, E., *Zen in the Art of Archery: Training the Mind and Body to Become One*, Arkana, London, 1988.

Heschel, A.J., *The Sabbath: Its Meaning for Modern Man*, Farrar, Straus and Giroux, New York, 1951.

Heyes, C., *Tripping: An Anthology of True-life Psychedelic Adventures*, Penguin Compass, New York, 2000.

Hofmann, A., *LSD My Problem Child: Reflections on Sacred Drugs, Mysticism and Science*, Jeremy Tarcher, Los Angelese; CA, 1983.

Holecek, A., *Dream Yoga: Illuminating Your Life Through Lucid Dreaming and the Tibetan Yogas of Sleep*, Sounds True, Louisville, CO, 2016.

Hölldobler, B. and Wilson E.O., *The Superorganism: The Beauty, Elegance, and Strangeness of Insect Societies*, W.W. Norton and Co., New York, NY, 2009.

Hopkins, E.W., 'Festivals and fasts (Hindu)', In: Hastings, J. (ed.), *Encyclopaedia of Religion and Ethics*, Vol. 5, Clark, Edinburgh, 1912.

Horowitz, M., *One Simple Idea: How the Lessons of Positive Thinking Can Transform Your Life*, Skyhorse Publishing, New York, 2014.

Howick, J., *Doctor You: Introducing the Hard Science of Self-Healing*, Coronet, London, 2017.

Huizinga, J., *Homo Ludens: A Study of the Play-Element in Culture*, Routledge and Kegan Paul, London, 1949.

Humphrey, N., *The Mind Made Flesh: Frontiers of Psychology and Evolution*, Oxford University Press, Oxford, 2002.

Humphrey, N., *Soul Dust: The Magic of Consciousness*, Quercus, London, 2011.

Humphrey, N. and Skoyles, J. , 'The evolutionary psychology of healing: a human success story', *Current Biology*, 2012, 22(17), 695–8.

Hutton, R., *The Stations of the Sun: A History of the Ritual Year in Britain*, Oxford University Press, Oxford, 1996.

Huxley, A., *The Doors of Perception and Heaven and Hell*, Flamingo, London, 1994.

Ibn 'Arabī, M., *Divine Sayings: Mishkāt al-anwār* (trans. Hirtenstein, S. and Notcutt, M.), Anqa Publishing, Oxford, 2004.

Jabr, F., 'Why your brain needs more downtime', *Scientific American*, 15 October 2013.

Jackson, S.A. and Csikszentmihalyi, M., *Flow in Sports*, Human Kinetics, Champaign, IL, 1999.

James, E.O., *Seasonal Feasts and Festivals*, Thames and Hudson, London, 1961.

James, W., *The Varieties of Religious Experience: A Study in Human Nature*, Fontana, London, 1960.

Jansen, K.L.R., 'Using ketamine to induce the near-death experience: mechanism of action and therapeutic potential', *Jahrbuch für Ethnomedizin und Bewußtseinsforschung*, 1996, 4, 55–81.

Jenner, N., 'Sun's fractal surprise could help fusion on earth', *New Scientist*, 30 April 2014.

Jeremiah, K., *Living Buddhas: The Self-Mummified Monks of Yamagata, Japan*, McFarland, Jefferson, NC, 2010.

Kamal, R.M., van Noorden, M.S., Franzek, E., Dijkstra, B.A.G., Loonen, A.J.M. and de Jong, C.A.J., 'The neurobiological mechanisms of gamma-hydroxybutyrate dependence and withdrawal and their clinical relevance: a review', *Neuropsychobiology*, 2016, 73, 65–80.

Karsh, E.B. and Turner, D.C., 'The human-cat relationship', In: Turner, D.C. and Bateson, P. (eds), *The Domestic Cat*, Cambridge University Press, Cambridge, 1988.

Kaelen, M., et al., 'LSD enhances the emotional response to music', *European Psychopharmacology*, 2015, 232: 3607–14.

Kerby, G. and Macdonald, D.W., 'Cat society and the consequences of colony size', In: Turner, D.C. and Bateson, P. (eds), *The Domestic Cat*, Cambridge University Press, Cambridge, 1988.

Kerndt, P.R., Naughton, J.L., Driscoll, C.E. and Loxterkamp, D.A., 'Fasting: the history, pathophysiology and complications', *Western Journal of Medicine*, 1982, 137, 379–99.

Kiecolt-Glaser, J.K., and Glaser, R., 'Depression and immune function:

central pathways to morbidity and mortality', *Journal of Psychosomatic Research*, 2002, 53, 873–6.

Koenig, H., King, D.E. and Carson, V.B., *Handbook of Religion and Health* (2nd ed.), Oxford University Press, Oxford, 2012.

Koenig, H., McCullough, M.E., and Larson, D.B., *Handbook of Religion and Health*, Oxford University Press, Oxford, 2001.

Krauss, L., *A Universe from Nothing: Why There is Something Rather Than Nothing*, Simon and Schuster, London, 2012.

Kretzmann, N., 'Philosophy of mind', In: Kretzmann, N. and Stump, E. (eds), *The Cambridge Companion to Aquinas*, Cambridge University Press, Cambridge, 1993.

Kuhn, T., *The Structure of Scientific Revolutions*, University of Chicago Press, Chicago, IL, 1962.

Leblanc, D., *Tithing: Test Me in This*, Thomas Nelson, Nashville, TN, 2010.

Lee, R.T., Kingstone, T., Roberts, L., Edwards, S., Soundy, A., Shah, P.R., Haque, M.S. and Singh, S., 'A pragmatic randomised controlled trial of healing therapy in a gastroenterology outpatient setting', *European Journal of Integrative Medicine* 2017, 9, 110–19.

Leskowitz, E., 'Group energies and sports team chemistry and fan energy', In: Leskowitz, E. (ed.), *Sports Energy and Consciousness: Awakening Human Potential Through Sport*, Create Space Publishing, North Charleston, SC, 2014.

Lindberg, J., Bjornerfeldt, S., Saetre, P., Svartberg, K., Seehuus, B., Bakken, M., Vila, C. and Jazin, E., 'Selection for tameness has changed brain gene expression in silver foxes', *Current Biology*, 2005, 22, 915–16.

Longo, V.D. and Mattson, M.P., 'Fasting: Molecular mechanisms and clinical applications', *Cell Metabolism*, 2014, 19, 181–92.

Lopez, S.J. and Snyder, C.R. (eds), *The Oxford Handbook of Positive Psychology* (2nd ed.), Oxford University Press, New York, 2009.

Lorimer, D. (ed.), 'Science and religion: A survey of spiritual practices and beliefs among European scientists, engineers and medical professionals', *Scientific and Medical Network Review*, 2017/1, 23–6.

Luhrmann, T.M., *When God Talks Back: Understanding the American Evangelical Relationship with God*, Vintage, New York, 2012.

Luke, D., *Otherworlds: Psychedelics and Exceptional Human Experience*, Muswell Hill Press, London, 2017.

Ly, C. et al., 'Psychedelics Promote Structural and Functional Neural Plasticity', *Cell Reports*, 2018, 23, 3170–82.

Lynch, J.J. and McCarthy, J.F., 'Social responding in dogs: heart rate changes to a person', *Psychophysiology*, 1969, 5, 389–93.

MacCulloch, J.A., 'Fasting', In: Hastings, J. (ed.), *Encyclopaedia of Religion and Ethics*, Vol. 5, Clark, Edinburgh, 1912.

McFarland, D. (ed.), *The Oxford Companion to Animal Behaviour*, Oxford University Press, Oxford, 1981.

McKenna, T., *Food of the Gods: The Search for the Original Tree of Knowledge; A Radical History of Plants, Drugs, and Human Evolution*, Bantam, London, 1993.

McPartland, J.M., Agraval, J., Gleeson, D., Heasman, K. and Glass, M., 'Cannabinoid receptors in invertebrates', *Journal of Evolutionary Biology*, 2005, 19, 366–73.

Mallinson, J. and Singleton, M., *Roots of Yoga*, Penguin, London, 2017.

Marchant, J., *Cure: A Journey into the Science of Mind Over Body*, Canongate, London, 2016.

Mascaro, J. (trans.), *The Upanishads*, Penguin, Harmondsworth, 1965.

Masson, J.M., *Dogs Never Lie About Love*, Cape, London, 1997.

May, E.C. and Marwaha, S.B. (eds), *Extrasensory Perception: Support, Skepticism, and Science*, Praeger, Santa Barbara, CA, 2015.

Meeusen, R., 'Exercise and the brain: insight in new therapeutic modalities', *Annals of Transplantation*, 2005, 10/4, 49–51.

Meggyesy, D., *Out of Their League*, University of Nebraska Press, Lincoln, NE, 2005.

Mikulas, W.L., 'Buddhism and Western psychology', *Journal of Consciousness Studies*, 2007, 14, 4–49.

Milne, J., *The Mystical Cosmos*, Temenos Academy, London, 2013.

Mithen, S., *The Prehistory of the Mind: A Search for the Origins of Art, Religion and Science*, Thames and Hudson, London, 1996.

Morell, V., 'The origin of dogs: running with the wolves', *Science*, 1997, 276, 1647–8.

Mohammad, Ghazi bin, *The Sacred Origin of Sports and Culture*, Fons Vitae Press, Louisville, KY, 1998.

Mohammad, Ghazi bin, *A Thinking Person's Guide to Islam*, White Thread Press, London, 2017.

Murphy, M., *The Future of the Body: Explorations into the Further Evolution of Human Nature*, Tarcher, Los Angeles, 1992.

Murphy, M., *Golf in the Kingdom*, Penguin Books, New York, 2011.

Murphy, M. and White, R., *The Psychic Side of Sports*, Addison-Wesley, Reading, MA, 1978.

Murphy, T., *Sacred Pathways: The Brain's Role in Religious and Mystic Experiences*, Todd Murphy, Amazon, 2015.

Nagel, T., *Mind and Cosmos: Why the Materialist, Neo-Darwinian Conception of Nature is Almost Certainly False*, Oxford University Press, Oxford, 2012.

Naranjo, C., *The One Quest: A Map of the Ways of Transformation*, Gateways Books, Nevada City, CA, 2006.

Neave, N. and Wolfson, S., 'Testosterone, territoriality and the "home advantage"', *Physiology and Behavior*, 2003, 78, 269–75.

Nestor, J., *Deep: Freediving, Renegade Science, and What the Ocean Tells Us About Ourselves*, Mariner Books, Houghton Mifflin Harcourt, Boston, MA, 2015.

Novak, M., *The Joy of Sports*, Basic Books, New York, 1976.

O'Donohue, J., *Benedictus: A Book of Blessings*, Bantam, London, 2007.

Ormerod, E., 'Pet programmes in prisons', *Society for Companion Animal Studies Journal*, 1996, 8(4), 1–3.

Otake, K., Shimai, S., Tanaka-Matsumi, J., Otsui, K. and Fredrickson, B.L., 'Happy people become happier through kindness: A counting kindnesses intervention', *Journal of Happiness Studies*, 2006, 7, 361–75.

Ott, J., *Pharmacotheon: Entheogenic Drugs, Their Plant Sources and History*, Natural Products Co., Kennewick, WA, 1993.

Owens, B., 'Drug development: the treasure chest', *Nature*, 2015, 525, S6–8.

Pagotto, U., Marsicano, G., Cota, D., Lutz, B. and Pasquali, R., 'The emerging role of the endocannabinoid system in endocrine regulation and energy balance', *Endocrine Reviews*, 2006, 27, 73–100.

Papineau, D., *Knowing the Score: What Sports Can Teach Us About Philosophy (and What Philosophy Can Teach Us About Sports)*, Basic Books, New York, 2017.

Partridge, E., *Origins*, Routledge and Kegan Paul, London, 1961.

Paul, E.S. and Serpell, J.A., 'Obtaining a new pet dog: effects on middle childhood children and their families', *Applied Animal Behaviour Science*, 1996, 47, 17–29.

Payne, P., *Martial Arts: The Spiritual Dimension*, Thames and Hudson, London, 1981.

Peale, N.V., *The Power of Positive Thinking for Young People*, Prentice-Hall, New York, NY, 1955.

Pérez-Manrique, A. and Gomila, A., 'The comparative study of empathy: sympathetic concern and empathic perspective-taking in non-human animals', *Biological Reviews*, 2018, 93, 248–69.

Peterson, C. and Park, N., 'Classifying and measuring strengths of character', In: Lopez, S.J. and Snyder, C.R. (eds), *The Oxford Handbook of Positive Psychology* (2nd ed.), Oxford University Press, New York, 2009.

Peterson, C. and Seligman, M.E.P., *Character strengths and Virtues: A Handbook and Classification*, Oxford University Press, New York, 2004.

Phear, D., 'A study of animal companionship in a day hospice', *Society for Companion Animal Studies Journal*, 1997, 9(1), 1–3.

Phillips, A. and Taylor, B., *On Kindness*, Penguin, London, 2009.

Bibliography

Plato (trans. Bury, R.G.), *Plato in Twelve Volumes: The Laws*, Harvard University Press, Cambridge, MA, 1967–8.

Pollan, M., *How to Change Your Mind: The New Science of Psychedelics*, Allen Lane, London, 2018.

Pontzer, H., 'Overview of hominin evolution', *Nature Education Knowledge*, 2012, 3(10), 8.

Pope Francis, *Laudato Si': On Care for Our Common Home*, Catholic Truth Society, London, 2015.

Popper, K. and Eccles, J.C., *The Self and its Brain*, Springer Verlag, Berlin, 1977.

Quickfall, J. and Crockford, D., 'Brain neuroimaging in cannabis use: A review', *Journal of Neuropsychiatry and Clinical Neurosciences*, 2006, 18, 318–32.

Rennie, A., 'The therapeutic relationship between animals and humans', *Society for Companion Animal Studies Journal*, 1997, 9(4), 1–4.

Ridley, M., *The Origins of Virtue*, Viking, London, 1996.

Rilling, J.K., Gutman, D.A., Zeh, T.R., Pagnoni, G., Berne, G.S. and Kilts, C.D., 'A neural basis for social cooperation', *Neuron*, 2002, 35, 395–405.

Rios, M.D. de and Janiger, O., *LSD: Spirituality and the Creative Process*, Park Street Press, Rochester, VT, 2003.

Roseman, L., Sereno, M.I. et al., 'LSD alters eyes-closed functional connectivity within the early visual cortex in a retinotopic fashion', *Human Brain Mapping*, 2016, 37, 3031–40.

Roud, S., *The English Year: A Month-by-Month Guide to the Nation's Customs and Festivals, from May Day to Mischief Night*, Penguin Books, London, 2006.

Rundle Clark, R.T., *Myth and Symbol in Ancient Egypt*, Thames and Hudson, London, 1959.

Saunders, N., *Ecstasy and the Dance Culture*, Saunders, London, 1995.

Schwartz, S.A. and Dossey, L., 'Nonlocality, intention and observer effect in healing studies: laying a foundation for the future', *Explore*, 2010, 6(5), 295–307.

Seligman, M., *Authentic Happiness: Using the New Positive Psychology to Realize Your Potential for Lasting Fulfilment*, Free Press, New York, NY, 2002.

Serpell, J., 'Best friend or worst enemy: cross-cultural variation in attitudes to the domestic dog', *Proceedings of the 1983 International Symposium of the Human-Pet Relationship*, Austrian Academy of Sciences, Vienna, 1983.

Serpell, J., 'Beneficial effects of pet ownership on some aspects of human health and behaviour', *Journal of the Royal Society of Medicine*, 1991, 84, 717–20.

Sessa, B., *The Psychedelic Renaissance: Reassessing the Role of Psychedelic Drugs in 21st Century Psychiatry and Society*, Muswell Hill Press, London. 2017.

Shanon, B., *The Antipodes of the Mind: Charting the Phenomenology of the Ayahuasca Experience*, Oxford University Press, New York, 2005.

Shanon, B., 'Biblical entheogens: A speculative hypothesis', *Time and Mind: The Journal of Archaeology, Consciousness and Culture*, 2008, 1, 51–74.

Sheldrake, R., *The Rebirth of Nature: The Greening of Science and God*, Century, London, 1990.

Sheldrake, R., *A New Science of Life: The Hypothesis of Formative Causation* (3rd ed.), Icon Books, London, 2009.

Sheldrake, R., *The Presence of the Past: Morphic Resonance and the Habits of Nature* (2nd ed.), Icon Books, London, 2011a.

Sheldrake, R., *Dogs that Know When Their Owners Are Coming Home, And Other Unexplained Powers of Animals* (2nd ed.), Three Rivers Press, New York, 2011b.

Sheldrake, R., *The Science Delusion: Freeing the Spirit of Enquiry*, Coronet, London, 2012 (In the US, published as *Science Set Free*).

Sheldrake, R., *The Sense of Being Stared At, and Other Unexplained Powers of Human Minds* (2nd ed.), Park Street Press, Rochester, VT, 2013.

Sheldrake, R., *Science and Spiritual Practices*, Coronet, London, 2017.

Sheldrake, R., Avraamides, L. and Novak, M., 'Sensing the sending of SMS messages: An automated test', *Explore: The Journal of Science and Healing*, 2009, 5, 272–6.

Sheldrake, R., McKenna, T. and Abraham, R., *The Evolutionary Mind: Conversations on Science, Imagination and Spirit*, Monkfish Books, Rhinebeck, NY, 2005.

Sheldrake, R. and Smart, P., 'A dog that seems to know when its owner is returning: preliminary investigations', *Journal of the Society for Psychical Research*, 1998, 62, 220–32.

Sheldrake, R. and Smart, P., 'A dog that seems to know when his owner is coming home: videotaped experiments and observations', *Journal of Scientific Exploration*, 2000a, 14, 233–55.

Sheldrake, R. and Smart, P., 'Testing a return-anticipating dog, Kane', *Anthrozoös*, 2000b, 13, 203–11.

Sheldrake, R. and Smart, P., 'Videotaped experiments on telephone telepathy', *Journal of Parapsychology*, 2003, 67, 147–66.

Sheldrake, R. and Smart, P., 'Testing for telepathy in connection with emails', *Perceptual and Motor Skills*, 2005, 101, 771–86.

Shulgin, A. and Shulgin, A., *PIHKAL: A Chemical Love Story*, Transform Press, Berkeley, CA, 1991.

Shulgin, A. and Shulgin, A., *TIHKAL: The Continuation*, Transform Press, Berkeley, CA, 1997.

Sinclair, U., *The Fasting Cure*, Mitchell Kennerley, New York, 1911.

Speth, J., Speth, C., Kaelen, M., Schloerscheidt, A.M., Feilding, A., Nutt, D.J. and Carhart-Harris, R. L., 'Decreased mental time travel to the past correlates with default-mode network disintegration under lysergic acid diethylamide', *Journal of Psychopharmacology*, 2016, 30, 344–53.

Steger, M.F., 'Meaning in life', In: Lopez, S.J. and Snyder, C.R. (eds), *Oxford Handbook of Positive Psychology* (2nd ed.), Oxford University Press, Oxford, 2009.

Stewart, W.K. and Fleming, L.W., 'Features of a successful therapeutic fast of 382 days' duration', *Postgraduate Medical Journal*, 1973, 49, 203–9.

Strassman, R., *DMT: The Spirit Molecule*, Park Street Press, Rochester, VT, 2001.

Strassman, R., *DMT and the Soul of Prophecy: A New Science of Spiritual Revelation in the Hebrew Bible*, Healing Arts Press, Rochester, VT, 2014.

Summerfield, H., 'Pets as therapy', *Society for Companion Animal Studies Journal*, 1996, 8(4), 9.

Taylor, C., *A Secular Age*, Harvard University Press, Cambridge, MA, 2007.

Thurston, H., *The Physical Phenomena of Mysticism*, Burns & Oates, London, 1952.

Tomasello, M., *Why We Cooperate*, Boston Review Books, Boston, MA, 2009.

Trubshaw, B., *Dream Incubation*, Heart of Albion Press, Loughborough, 2017.

Trut, I.N., Plyusnina, I.Z. and Oskina, I.N., 'An experiment on fox domestication and debatable issues of evolution of the dog', *Genetika*, 2004, 40, 794–807.

Turnbull, C., *The Forest People*, The Bodley Head, London, 2015.

Vitebsky, P., *Reindeer People: Living with Animals and Spirits in Siberia*, Harper Perennial, London, 2005.

Waal, F. de, *Primates and Philosophers: How Morality Evolved*, Princeton University Press, Princeton, NJ 2009.

Warneken, F. and Tomasello, M., 'Helping and cooperation at 14 months of age', *Infancy*, 2007, 11, 271–94.

Webster, H., 'Sabbath (Primitive)', In: Hastings, J. (ed.), *Encyclopaedia of Religion and Ethics*, Clark, Edinburgh, Vol. 10, 1918.

Whistler, L., *The English Festivals*, Heinemann, London, 1947.

Whitehouse, H., 'Dying for the group: Towards a general theory of extreme self-sacrifice', *Behavioral and Brain Sciences*, 2018 (February), doi:10.1017/S0140525X18000249.

Wilson E.O., *Sociobiology*, Harvard University Press, Cambridge, MA, 1980.

Wiseman, E., 'The cult of being kind', *The Observer Magazine*, 1 April 2018, 12–15.

Wood, R.I. and Stanton, S.J., 'Testosterone and sport: current perspectives', *Hormones and Behavior*, 2012, 61, 147–55.

Wroe, A., *Being Shelley: The Poet's Search for Himself*, Vintage Books, London, 2007.

Xu B., Ji, L., Lin, L. and Xu, F., 'The influence of swimming on learning-memory of rats and on DA concentration in rat's brain', *Chinese Journal of Sports Medicine*, 2004, 23, 261–5.

Zaehner, R., *Mysticism Sacred and Profane: An Inquiry into Some Varieties of Praeternatural Experience*, Oxford University Press, Oxford, 1957.

Index

317

Index

Index

Index

Index